Lectures on Exceptional Lie Groups

Chicago Lectures in Mathematics Series
Robert J. Zimmer, series editor
J. Peter May, Spencer J. Bloch, Norman R. Lebovits,
William Fulton, and Carlos Kenig, editors

Other *Chicago Lectures in Mathematics* titles available from the University of Chicago Press:

Simplicial Objects in Algebraic Topology, by J. Peter May (1967)
Fields and Rings, Second Edition, by Irving Kaplansky (1969, 1972)
Lie Algebras and Locally Compact Groups, by Irving Kaplansky (1971)
Several Complex Variables, by Raghavan Narasimhan (1971)
Torsion-Free Modules, by Eben Matlis (1973)
Stable Homotopy and Generalised Homology, by J. F. Adams(1974)
Rings with Involution, by I. N. Herstein (1976)
Theory of Unitary Group Representation, by George V. Mackey (1976)
Infinite-Dimensional Optimization and Convexity, by Ivar Ekeland and Thomas Turnbull(1983)
Navier-Stokes Equations, by Peter Constantin and Ciprian Foias (1988)
Essential Results of Functional Analysis, by Robert J. Zimmer (1990)
Fuchsian Groups, by Svetlana Katok (1992)
Unstable Modules over the Steenrod Algebra and Sullivan's Fixed Point Set Conjecture, by Lionel Schwartz (1994)
Topological Classification of Stratified Spaces, by Shmuel Weinberger (1994)

J. F. Adams

LECTURES ON EXCEPTIONAL LIE GROUPS

Edited by Zafer Mahmud and Mamoru Mimura

The University of Chicago Press
Chicago and London

J. F. Adams (1930–1989) was Lowndean Professor of Astronomy and Geometry at the University of Cambridge. Zafer Mahmud is associate professor of mathematics at Kuwait University. Mamoru Mimura is professor of mathematics at Okayama University.

The University of Chicago Press, Chicago 60637
The University of Chicago Press Ltd., London

Printed in the United States of America
04 03 02 01 00 99 98 97 96 1 2 3 4 5
ISBN 0–226–00526–7 (cloth)
0–226–00527–5 (paper)

Library of Congress Cataloging–in–Publication Data

Adams, J. Frank (John Frank)
Lectures on exceptional Lie groups / J.F. Adams ; edited by Zafer Mahmud and Mamoru Mimura.
p. cm. —(Chicago lectures in mathematics series)
Includes bibliographical references.
ISBN 0-226-00526-7 (cloth : alk. paper). —ISBN 0-226-00527-5 (ppk. : alk. paper)
1. Lie groups. I. Mahmud, Zafer. II. Mimura, M. (Mamoru). 1938–
III. Title. IV. Series: Chicago lectures in mathematics.
QA387.A33 1996
512'.55—dc20
96-21456
CIP

The paper used in this publication meets the minimum requirements of the American National Standard for Information Science—Permanence of Paper for Printed Library Materials, ANSI Z39.48-1984.

Contents/Summary

Summary of Constructions

Construction (possibly) using E_8	Direct construction
$G_2 =$ subgroup of $Spin(7)$ fixing $z \in S^7$	$Aut_{\mathbf{R}}(\mathbf{O})$
$F_4 = C^0_{E_8}(G_2), G_2 \to Spin(7) \to Spin(16) \to E_8$	$Aut(J)$, $J =$ algebra
$E_6 = C^0_{E_8}(SU(3)), SU(3) \to Spin(16) \to E_8$	$Aut_0(W, W^*, \times)$
$E_7 = C^0_{E_8}(SU(2)), SU(2) \to Spin(16) \to E_8$	$Aut_0(W, \langle\ ,\ \rangle, f), f : W^{\otimes 4} \to \mathbf{C}$
E_8	$Aut(L(Spin(16) \oplus \Delta^+))$

Foreword

Frank Adams was long intrigued by the exceptional Lie groups, and he wrote several papers about them ([Ad1, Ad2, Ad3] in the bibliography). He worked out a connected narrative describing their construction, their representations, and their interconnections and connections with the classical Lie groups in several lecture series given at Cambridge University. He often contemplated turning his lecture notes into a book, but unfortunately this was prevented by his untimely death. At the Memorial Symposium that was held in his honor at the University of Manchester in August of 1990, Professors Zafer Mahmud and Mamoru Mimura graciously agreed to edit the several versions of Adams' lecture notes, with the further aid of the notes taken by Wilson Sutherland and Adams' student, John Greenlees. Professors Mahmud and Mimura have carefully carried out this task. Since it was Adams' custom to write out lecture notes in unusual detail, the resulting volume faithfully reproduces his own words, but it is to be understood that Mahmud and Mimura were working from lecture notes that Adams himself would undoubtedly have reworked before publication. Nevertheless, we believe that the resulting volume is true to Adams' spirit and will be of considerable value to all mathematicians who wish to learn about those wonderful curiosities, the exceptional Lie groups. The notes themselves give excellent summaries of the required background in Lie group theory, details of which may be found in Adams' own volume *Lectures on Lie Groups*, which is also available from the University of Chicago Press.

J. Peter May

Acknowledgments

Zafer Mahmud wishes to thank Professors S. J. Patterson and Larry Smith of the Mathematics Institute, Göttingen, for very generous hospitality and facilities during the year 1990-91. Without this support, he could not have participated in this project.

Introduction

The object of this work is to study the exceptional Lie groups. These may be regarded as curiosities of Nature though all right-thinking mathematicians are glad they exist. In particular, some theoretical physicists are beginning to wonder if the exceptional groups turn up in quantum theory. In algebraic topology, their homology and cohomology rings provide examples of Hopf algebras to be reckoned with in formulating general structure theorems. Again, people interested in finite groups are interested in finite groups related to these Lie groups.

The book should be seen as a sequel to [1]. The aim is to give concrete and explicit constructions of the exceptional Lie algebras and Lie groups with the most interesting low dimensional representations. We will not give the general theory or the most conceptual definitions. Most of what we need can be found in [1].

Before we get to the five exceptional groups, there are other matters we should mention first: the groups $Spin(n)$ and their representations Δ (n odd), $\Delta^{\pm}$ (n even). This must be preceded by something about Clifford algebras.

To begin with, we have general chapters about Lie groups. This is not because we are going to need theorems: you can always study an explicit concrete special case without the aid of general theory. But we need language and standard constructions, for example: passing from a Lie group to its Lie algebra. More specifically, we need methods to recognise that the exceptional groups we construct are not isomorphic to any of the classical groups.

The prerequisites for this book are [1] or a third year undergraduate course in differential geometry (in a British University), linear algebra (tensor products and *Hom* functors etc., such as in [La]) and rudiments of topology. (The US equivalent prerequisites would be advanced calculus, linear algebra, differential geometry, basic topology and a first course in abstract algebra.) Thus, where a map or space ought to be smooth, we hardly ever check that it actually is, leaving it to the reader to see that the map or space in question is obviously smooth. Apart from these prerequisites, we supply detailed proofs throughout and make all constructions *ab initio*.

The first six chapters are preparatory. In many of our constructions it is possible to work simultaneously over the reals and complex numbers. If it is not explicitly stated, the context should make clear the ground field over which we are working. We construct G_2 in Chapter 5, the Lie algebra E_8 in Chapter 6 and the Lie group E_8 in Chapter 7. In Chapter 8, F_4, E_6 and E_7 are constructed as the connected components of the centralisers of various subgroups of E_8, and certain standard representations identified. The Dynkin diagrams are given in Chapter 9, the Weyl group of E_8 discussed in Chapter 10, and properties of the standard representations of E_6 and E_7 derived in Chapter 11. Chapters 12, 13, 14, and 16 are largely independent of the rest of the book and give constructions of F_4, E_6, E_7 and their representations, not using the group E_8. The idea here is to construct a vector space equipped with extra structure. This space turns out to be the representation constructed previously and the groups appear as the group of automorphisms preserving all the structure in sight. In particular, F_4 appears as the automorphism group of an exceptional Jordan algebra (there is a brief appendix listing simple properties of Jordan algebras). In similar vein, in Chapter 15 we construct the algebra of Cayley numbers and identify G_2 as the automorphism group. The reader can get an idea of these constructions by consulting the CONTENTS/SUMMARY. The end of a proof is indicated thus: $\square$.

Chapter 1

Definitions, examples and matrix groups

We assume familiarity with the basic theory of compact Lie groups and Lie algebras to be found in [1] (other references are [7], [9]). In this Chapter we establish the notation that will be used in the sequel.

A *Lie group* G is a group which is also a smooth manifold such that the maps $G \times G \to G$, $(g,h) \mapsto gh$ and $G \to G$, $g \mapsto g^{-1}$ are smooth. A *homomorphism* $\theta : G \to H$ of Lie groups is a homomorphism of groups which is also a smooth map, [1, p. 3]. A *representation* V of G is a representation in the sense of algebra, [1, p. 23, definitions 3.1], such that the map $G \times V \to V$, $(g,v) \mapsto gv$ is smooth. Let $\Lambda = \mathbf{R}$, $\mathbf{C}$ or $\mathbf{H}$ be respectively the reals, the complex numbers or the quaternions and V a finite dimensional right Λ-module. Then, by choosing a basis for V, we get isomorphisms $V \cong \Lambda^n$ for some integer n and

$$\begin{aligned} End(V) &= End_\Lambda(V) = Hom_\Lambda(V,V) \\ &\cong M_n(\Lambda) = \{\, n \times n \text{ matrices with entries in } \Lambda \,\}. \end{aligned}$$

The general linear group of V,

$$GL(V) \;= Aut(V) = \{\, A \in End(V) \mid \exists\, A^{-1} \in End(V) \,\},$$

is a group, and an open subset of $End(V)$ (= a finite dimensional vector space over Λ). This ensures that it is a smooth manifold. The product and inverse map in $GL(V)$ are smooth, so $GL(V)$ is a Lie group. If the dimension of V over Λ is n, we will write $GL(n,\Lambda)$ for $GL(V)$. We can choose on V a Hermitian form $(\ ,\ )$ such that

$$\begin{aligned} &(v, w\lambda) = (v,w)\lambda \quad \text{for } \lambda \in \Lambda,\ v,w \in V, \text{ and} \\ &(w,v) = \overline{(v,w)}, \end{aligned}$$

and then consider such subgroups of $GL(V)$ as

$$\{\, g \in GL(n,\Lambda) \mid (gv, gw) = (v,w), \quad \forall\, v,w \in V \,\}.$$

Here $\overline{(v,w)}$ denotes the conjugate in Λ, see Example 2 at the end of Chapter 2.

By choosing a basis in V and defining $(x,y) = \bar{x}^T y$, we see that for all $v \in V$, $(v,v) \in \mathbf{R}$, $(v,v) \geq 0$ with $=$ if and only if $v = 0$.

The subgroups $\{ A \in GL(n,\Lambda) \mid \bar{A}^T A = I \}$ are denoted $O(n)$, $U(n)$, $Sp(n)$ respectively for $\Lambda = \mathbf{R}$, $\mathbf{C}$, $\mathbf{H}$. For $\Lambda = \mathbf{R}$ or $\mathbf{C}$, we also get subgroups $\{ A \in GL(n,\Lambda) \mid \det A = 1 \}$ denoted $SL(n,\mathbf{R})$, $SL(n,\mathbf{C})$ (called the *special linear groups*) and

$$SO(n) = SL(n,\mathbf{R}) \cap O(n), \quad SU(n) = SL(n,\mathbf{C}) \cap U(n).$$

All these groups are collectively called *classical groups* (and are in fact Lie groups), see [7, pp. 1–24].

Definition. *A* matrix group *is a closed subgroup of* $GL(V)$ *for some* V.

Infinitesimal methods.

The set of all tangent vectors at a point P of a manifold X is denoted X_P, and if X is modelled on a vector space V then X_P will be a vector space isomorphic to V. For each of the matrix groups we have met, we list the space of tangent vectors at the identity.

Proposition. *Suppose* $G = O(n)$, $U(n)$ *or* $Sp(n)$. *Then* $B \in G_1$ *if and only if* $\bar{B}^T + B = 0$. *For the groups* $G = SL(n,\mathbf{R}), SL(n,\mathbf{C})$, $B \in G_1$ *if and only if* $Tr(B)$, *the trace of* B, *is zero.*

□

Next we note that there is a Lie algebra structure on $G_1 = L(G)$ and that L is a functor from the category of Lie groups and homomorphisms to the category of Lie algebras and Lie algebra homomorphisms and that a representation of G gives a representation of $L(G)$.

Theorem 1.2. (i) *For any Lie group* G, *the space* G_1 *of tangent vectors to* G *at* 1 *becomes a Lie algebra* $L(G)$ *over* $\mathbf{R}$.

(ii) *For any matrix group* $G \subset GL(V)$, *the bracket* $[\gamma,\delta]$ *in* $L(G)$ *is the commutator* $\gamma\delta - \delta\gamma \in End(V)$.

(iii) *If* $\theta : G \to H$ *is a homomorphism of Lie groups, the induced map* $\theta_1 : G_1 \to H_1$ *becomes a homomorphism* $L(\theta) : L(G) \to L(H)$ *of Lie algebras.*

(iv) *For any representation* V *of* G, V *becomes a representation of* $L(G)$.

(v) *For a matrix group* $G \subset GL(V)$ *acting on* V *as given, the action of* $L(G) \subset End(V)$ *is the usual action of* $End(V)$ *on* V.

(vi) *If* H *acts on* V *and we are given a homomorphism* $\theta : G \to H$ *of Lie groups, so that* G *acts on* V, *the resulting action of* $L(G)$ *on* V *is* $\gamma \cdot v = (\theta_1\gamma)v$ *where* $\theta_1 = L(\theta)$ *and* $\gamma \in L(G)$.

A vector space V over $\Lambda' = \mathbf{R}$, $\mathbf{C}$ or $\mathbf{H}$, $\Lambda \subset \Lambda'$, is a *representation* of a Lie algebra A if we are given products $xv \in V$ for all $x \in A$, $v \in V$, Λ-linear in x, Λ'-linear in v and satisfying $[x,y]v = x(yv) - y(xv)$.

G acts on itself by conjugation: for all $g \in G$, we have $i_g : G \to G$, $i_g(h) = ghg^{-1}$, which gives $(i_g)_1 : G_1 \to G_1$, so G acts on $G_1 = L(G)$. This is the *adjoint representation*, $Ad : G \to Aut(L(G))$.

References for Theorem 1.2 are [11, pp. 421–432] and [4, Chapter 3, pp. 120–165].

Definition. *A Lie algebra is* simple *if it is of dimension more than one and has no proper ideals; a Lie algebra is* semisimple *if it is has no abelian ideal. A Lie group is* simple (semisimple) *if its Lie algebra is simple (semisimple).*

For example, $SU(n), n > 1$, and its Lie algebra are simple; over $\mathbf{R}$ or $\mathbf{C}$ a Lie algebra is semisimple if and only if it is the direct sum of simple algebras.

Of course, $L(G)$ only sees a neighbourhood of 1.

Examples. (i) $SO(n)$ is the identity component of $O(n)$ (which has two components) and

$$L(SO(n)) = \{\text{real skew symmetric } n \times n \text{ matrices}\} = L(O(n)).$$

(ii) We note that $U(1) = \{ z \mid z\bar{z} = 1 \} = S^1 \subset \mathbf{C}$ and the exponential map $\mathbf{R} \to S^1$, $t \mapsto e^{2\pi i t}$, induces an isomorphism of Lie algebras (of $\mathbf{R}$ and S^1) both of which are of dimension 1 and have zero brackets. Of course, the groups $\mathbf{R}$ and S^1 are not isomorphic.

(iii) The quotient map $SU(n) \to PSU(n) = SU(n)/\{ wI \mid w^n = 1 \}$ gives an isomorphism of Lie algebras. □

Even so, $L(G)$ *does* determine G locally, [Ha, p. 71].

Representation theory of compact groups.

(See [1, Chapter 3, pp. 27–78], [3, Chapter 6, pp. 171–213].)

The groups $O(n), U(n), Sp(n), SO(n), SU(n), PSU(n)$ are compact (whilst $GL(n,\Lambda)$, and $SL(n,\Lambda)$, $n \neq 1$, are not).

Most of the representation theory of finite groups goes over to compact groups with integration a substitute for finite sums. Thus for any real valued continuous function f on G we can define $\int_{g \in G} f(g)$ and it has the properties expected of $\frac{1}{|G|} \sum_{g \in G} f(g)$, [1, p. 33].

We shall focus on complex representations. For certainly, a representation V over $\mathbf{H}$ is a representation over $\mathbf{C}$. This representation over $\mathbf{C}$, together with a conjugate linear structure G-map $j : V \to V$ such that $j^2 = -1$, $ij = -ji$ gives the original $\mathbf{H}$-representation. A representation V over $\mathbf{R}$ gives $V \otimes_{\mathbf{R}} \mathbf{C}$ and this carries a conjugate linear structure map $j : v \otimes z \mapsto v \otimes \bar{z}$ such that $j^2 = 1$ and we can regard V as the $+1$ eigenspace of j (or the -1 eigenspace). See [1, pp. 24–33, 44–45].

Weights and characters.

One of the benefits of working over $\mathbf{C}$ is that every irreducible representation of a compact abelian group is 1-dimensional. For example, let T be the torus $\overset{d}{\underset{1}{\times}} S^1$. Another way to regard T is that since it is a connected compact abelian group, $\exp : L(T) \to T$ is a homomorphism, [1, p. 15, proof of Theorem 2.19], so $T \cong L(T)/\ker\exp$. Write $\Gamma = \ker\exp$; then Γ is a discrete subgroup of $L(T)$, called the *integer lattice* of $L(T)$ and choosing a basis gives $L(T) \cong \mathbf{R}^d$ for some d, [1, p. 15] and

$$T \cong L(T)/\Gamma \cong \mathbf{R}^d/\mathbf{Z}^d \cong (\mathbf{R}/\mathbf{Z})^d.$$

Homomorphisms $\theta : T \to T'$ are easily described. We need only check, for a linear map $\varphi : L(T) \to L(T')$, that $\varphi(\Gamma) \subset \Gamma'$, and if so, then $\theta = \bar{\varphi} : L(T)/\Gamma \to L(T')/\Gamma'$. All continuous homomorphisms arise in this way, cf. [1, p. 76, Proposition 3.74], and all 1-dimensional representations of T arise from linear maps $\varphi : L(T) \to L(U(1))$, [1, Corollary 3.75]. Given a representation V of T, there are linear maps $\varphi : L(T) \to L(U(1))$ such that V decomposes as a direct sum of non-zero sub-representations V_φ where $L(T)$ acts on V_φ by $\tau(x) = \varphi(\tau)x$ ($\tau \in L(T), x \in V_\varphi$). These linear maps φ on $L(T)$ are called the *weights* of V, and the dimension of V_φ the *multiplicity* of φ [1, pp. 75–77]. The space of tangent vectors to $S^1 \subset \mathbf{C}$ exists of course, and it is $i\mathbf{R}$. We will take $\Gamma = 2\pi i\,\mathbf{Z}$. (If we insisted on integral coordinates, we would get many useless factors of $2\pi i$.)

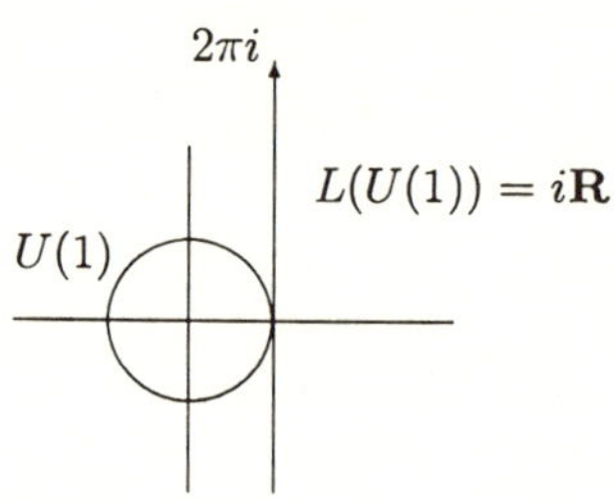

Fig. 1

Suppose V is a representation of G over $\mathbf{C}$. Then the *character*, $\chi_V : G \to \mathbf{C}$, of V is $\chi_V(g) = Tr_{\mathbf{C}}(g : V \to V)$. Characters are class functions, that is,

$$\chi_V(ghg^{-1}) = \chi_V(h)$$

for all $g, h \in G$. Also

$$\chi_{V\oplus W}(g) = \chi_V(g) + \chi_W(g), \quad \chi_{V\otimes W}(g) = \chi_V(g) \cdot \chi_W(g).$$

Theorem 1.3. *If* $\chi_V = \chi_W$ *then* $V \cong W$.
See [1, pp. 46–52, especially Proposition 3.37].

Sketch of the classification of compact Lie groups.

This involves constructing the Dynkin diagram of a compact Lie group and for this we need

Definition. A maximal torus *in a compact connected Lie group* G *is a subgroup* T *which is*

(i) *a torus* *and* (ii) *maximal,*

that is, if $T \subset T' \subset G$, T' *a torus, then* $T' = T$.

A general reference for maximal tori is [1, pp. 79–100].

Examples 1.4. See [1, examples 4.16–4.20].

(i) In $U(n)$ consider $diag(e^{2\pi i x_1}, \dots, e^{2\pi i x_n})$, the diagonal matrices with diagonal entries $e^{2\pi i x_j}$ for $x_j \in \mathbf{R}$. These form a maximal torus in $U(n)$.

(ii) Since $\mathbf{C} \subset \mathbf{H}$, the matrices of (i) are in $Sp(n)$, and they form a torus in $Sp(n)$. Since $\mathbf{C}^n$ can be regarded as $\mathbf{R}^{2n}$, we get an embedding $U(n) \to SO(2n)$ and we can again take the corresponding matrices, namely

$$\begin{pmatrix} \cos 2\pi x_1 & -\sin 2\pi x_1 & & & \\ \sin 2\pi x_1 & \cos 2\pi x_1 & & 0 & \\ & & \ddots & & \\ & 0 & & \cos 2\pi x_n & -\sin 2\pi x_n \\ & & & \sin 2\pi x_n & \cos 2\pi x_n \end{pmatrix}$$

These will form a torus in $SO(2n)$.

(iii) We can embed $\mathbf{R}^{2n}$ in $\mathbf{R}^{2n+1}$ and thus $SO(2n)$ in $SO(2n+1)$, where $A \mapsto \left(\begin{smallmatrix} A & 0 \\ 0 & 1 \end{smallmatrix}\right)$ and so get a corresponding torus in $SO(2n+1)$.

(iv) In $SU(n)$ we take the matrices of (i) subject to $\sum x_i = 0$ to get a maximal torus.

These tori are in fact all maximal. For example, in $U(n)$ any matrix X which commutes with all diagonal matrices must be diagonal, so in T. Thus

T is maximal among all abelian subgroups, connected or not. The existence of maximal tori is easy to prove in any Lie group. It is more substantial to prove that they have the properties one wants; these properties mirror the behaviour of diagonal matrices in $U(n)$.

Theorem 1.5. *Let $T \subset G$ be a maximal torus of a compact, connected, Lie group G. Then any $g \in G$ is conjugate to some element of T. That is, there exists $t \in T$, $h \in G$ such that $g = hth^{-1}$.*

See the proof in [1, pp. 89-92]. □

Corollary 1.6. *If V, W are representations of a compact, connected, Lie group G and $\chi_V|T = \chi_W|T$, then $\chi_V = \chi_W$, so $V \cong W$.*

Proof. The torus provides at least one representative from each conjugacy class, see [1, p. 95]. □

If V is a representation of G, the *weights* of V are, by definition, the weights of $V|T$. The weights (together with their multiplicities) determine V up to equivalence.

Corollary 1.7. *Given two maximal tori T, T' in a compact, connected, Lie group G, there exists an inner automorphism of G taking T to T'.*

Proof. [1, pp. 92–93]. □

It follows from Corollary 1.7 that any property of G defined by reference to a maximal torus T is independent of the choice of T.

Example: (Definition.) *The* rank *of a compact, connected, Lie group G is the dimension of a maximal torus of G.* (We will usually write $\ell = \text{rank G}$.)

Definition 1.8. *The* Weyl group W *of a compact, connected, Lie group G is the group of those automorphisms of a maximal torus which are given by inner automorphisms of G.*

Example. In $U(2)$ conjugation by $\left(\begin{smallmatrix} 0 & -1 \\ 1 & 0 \end{smallmatrix}\right)$ is an element of W and

$$\begin{pmatrix} 0 & -1 \\ 1 & 0 \end{pmatrix} \begin{pmatrix} e^{2\pi i x_1} & 0 \\ 0 & e^{2\pi i x_2} \end{pmatrix} \begin{pmatrix} 0 & 1 \\ -1 & 0 \end{pmatrix} = \begin{pmatrix} e^{2\pi i x_2} & 0 \\ 0 & e^{2\pi i x_1} \end{pmatrix}.$$

More generally, the Weyl groups of our matrix groups are as follows [1, pp. 114-116]:

$U(n)$ and $SU(n)$: $W = \Sigma_n$ (any permutation of $x_1, \dots, x_n$).
$Sp(n)$ and $SO(2n+1)$: $W =$ the group generated by all permutations of $x_1, \dots, x_n$ and all sign changes of x_i.
$SO(2n)$: $W =$ the group generated by all permutations of $x_1, \dots, x_n$ and an even number of sign changes of x_i.

□

We now have enough invariants to settle some questions. For example, if we have a Lie group (such as E_6, see Chapter 8) of dimension 78 and $|W| = 2^7 \cdot 3^4 \cdot 5$ then we know that $G \not\cong Sp(6)$, since $|W(Sp(6))| = 2^6 \cdot 6! = 2^{10} \cdot 3^2 \cdot 5$.

However, we need to go further.

Suppose G is a Lie group and $T \subset G$ a torus (not necessarily maximal). Then G acts on $L(G)$ via the adjoint representation, so T acts on $L(G)$ by restriction and $L(G) \otimes \mathbf{C}$ splits as a sum of 1-dimensional representations of T, with T acting trivially on $L(T) \subset L(G)$. Thus the trivial 1-dimensional representation occurs at least $d = \dim T$ times.

Proposition 1.9. *If G is compact, then T is maximal $\iff$ the trivial 1-dimensional representation occurs exactly d times.*

Proof. [1, p. 83].

From now on suppose T is maximal and put $d = \ell$.

Definition. *The* roots *of a compact, connected, Lie group G are the weights of the adjoint representation, excluding 0 (which occurs, by Proposition 1.9, ℓ times).*

The roots are thus $\mathbf{R}$-linear functions on $L(T)$, that is, elements of $L(T)^*$. Since the adjoint representation of G is real, the 1-dimensional summands of $L(G) \otimes \mathbf{C}$ occur in conjugate pairs and the roots occur in pairs $\pm\theta$.

Examples. (See [1, pp. 83–89].) We list the weights of $L(G) \otimes \mathbf{C}|T$ for the matrix groups G and maximal tori we have described in Examples 1.4.

$U(n)$: weights are $\pm(x_i - x_j)$ $1 \le i < j \le n$, $\quad$ 0 n times.

Proof. First note that $L(U(n)) \otimes \mathbf{C} \cong L(GL(n, \mathbf{C})) \cong End(\mathbf{C}^n)$ since $B \mapsto -\bar{B}^T$ is a conjugate linear structure map and its $+1$ eigenspace is $L(U(n))$.

Take the basis $\{e_j\}$ for $\mathbf{C}^n$, where e_j is the transpose of $(0, \dots, 0, 1, 0, \dots, 0)$ with "1" in the j-th place. For $i < j$, define linear maps $\theta_{ij} \in End(\mathbf{C}^n)$ by $\theta_{ij}(e_j) = e_i, \theta_{ij}(e_k) = 0, k \neq j$; the matrix of θ_{ij} has a 1 in the ij-th place and zeros elsewhere. The θ_{ij} are eigenvectors of the action of T with eigenvalues $\exp(2\pi\sqrt{-1}(x_i - x_j))$, so $x_i - x_j$ are eigenvalues for the action of $L(T)$ on $L(U(n))$.

(We are taking $L(T)$ to be the diagonal matrices $diag(\sqrt{-1}\,y_1, \ldots, \sqrt{-1}\,y_n), y_j \in \mathbf{R}$, and $x_i \in L(T)^*$ is given by $x_i(diag(\sqrt{-1}\,y_1, \ldots, \sqrt{-1}\,y_n)) = \sqrt{-1}\,y_i$.)

□

$SU(n)$	:	$\pm(x_i - x_j)$	$1 \le i < j \le n,$	0	$(n-1)$ times.
$SO(2n)$	:	$\pm x_i \pm x_j$	$1 \le i < j \le n,$	0	n times.
$SO(2n+1)$	:	$\pm x_i \pm x_j$	$1 \le i < j \le n,$	0	n times.
		$\pm x_i$	$1 \le i \le n.$		
$Sp(n)$	:	$\pm x_i \pm x_j$	$1 \le i < j \le n,$	0	n times.
		$\pm 2x_i$	$1 \le i \le n.$		

□

The Weyl group W acts on $L(T)$ and permutes roots. If we regard the roots as elements of $L(T)^*$, they form a configuration with great symmetry and very distinctive properties, [1, Chapter 5, pp. 101–141]. The Dynkin diagram encodes this configuration, as we proceed to describe. We may choose on $L(T)^*$ a positive definite inner product invariant under W, so that we can talk about *lengths* of roots, *angle* between roots and so on. For each pair of roots $\pm\theta$, $\ker\theta = \ker(-\theta)$ is a hyperplane in $L(T)$ called a *root plane.* Conversely, it can be shown that each root plane comes from only one pair of roots, $\pm\theta$, [1, pp. 105–108]. The root planes form a figure in $L(T)$ called the *(infinitesimal) Stiefel diagram.*

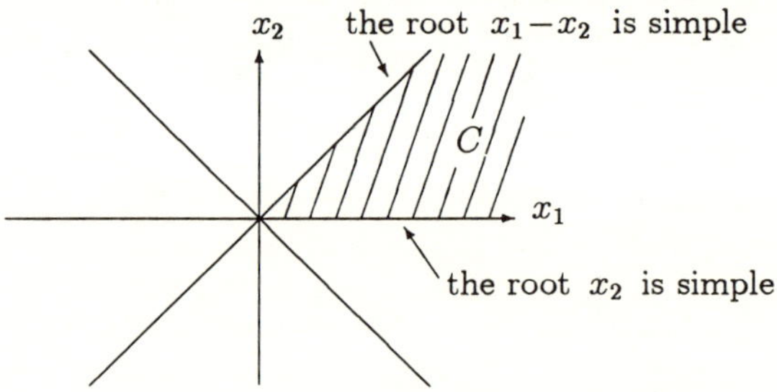

Fig. 2 The Stiefel diagram of $SO(5)$

The root planes divide $L(T)$ into convex open sets called *Weyl chambers* and the Weyl group permutes the Weyl chambers in a way which is simply transitive, [1, pp. 110–113]. We choose one and call it the *fundamental Weyl chamber* (FWC), C, shaded in Fig. 2. A root θ is *positive* (respectively *negative*) if $\theta > 0$ on C $(\theta < 0)$. A positive root is *simple* if it defines a wall of C. The *Dynkin diagram* has one node for each simple root (i.e. for each wall of the fundamental Weyl chamber). These nodes are joined by the following number of bonds:

$$\begin{cases} 0 & \text{if the roots are orthogonal} \\ 1 & \quad \dots \qquad\qquad \text{at } 120^\circ \\ 2 & \quad \dots \qquad\qquad \text{at } 135^\circ \\ 3 & \quad \dots \qquad\qquad \text{at } 150^\circ, \end{cases}$$

and these are the only possibilities, [1, pp. 118–121].

Examples 1.10. T, a torus, has the empty Dynkin diagram.

$U(n)$. Take C to have $x_1 < x_2 < \cdots < x_n$.

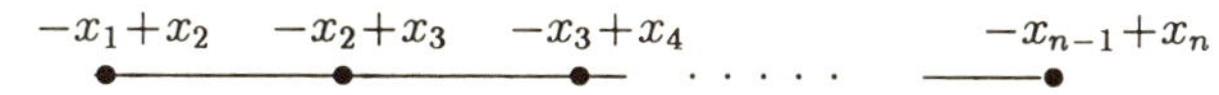

We take the usual inner product of $\mathbf{R}^n$.

$SU(n)$ has the same Dynkin diagram as $U(n)$, which is traditionally labeled A_{n-1}.

$SO(2n)$. Take C to have $-x_2 < x_1 < x_2 < x_3 < \cdots < x_n$. (D_n)
Then the diagram is

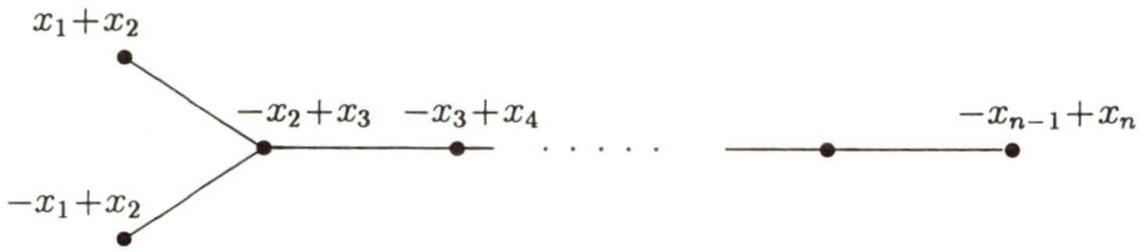

$SO(2n+1)$. Take C to have $0 < x_1 < x_2 < x_3 < \cdots < x_{n-1} < x_n$. (B_n)

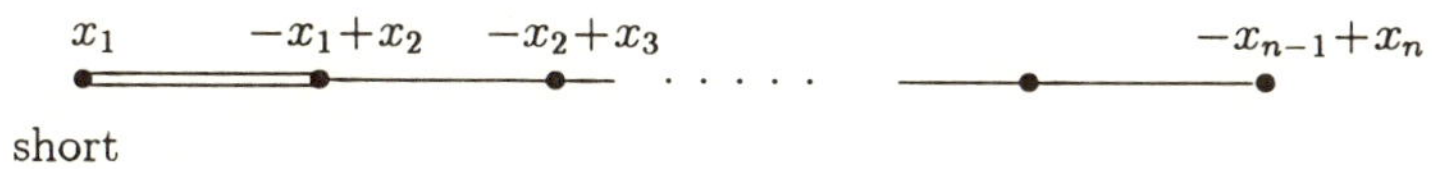

$Sp(n)$. Take C to have $0 < x_1 < x_2 < x_3 < \cdots < x_n$. (C_n)

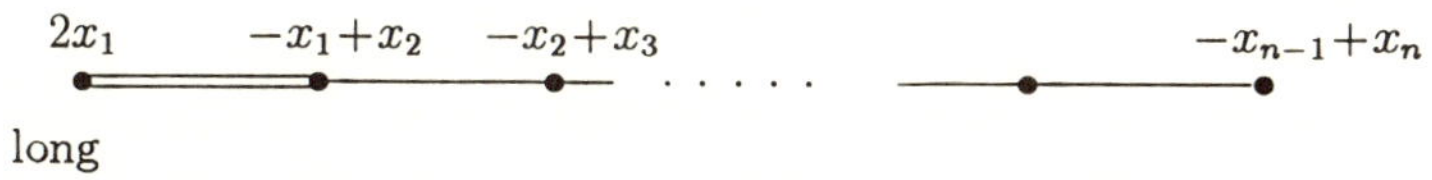

Note: "short" (respectively "long") means $x_1 \cdot x_1 = 1 < 2$ $(2x_1 \cdot 2x_1 = 4 > 2)$.

We now have a rule which associates to a compact Lie group a Dynkin diagram. However, as yet we only hope to distinguish between Lie groups with different Lie algebras over $\mathbf{R}$. To see this, note that we can recognise $L(T)$ in $L(G)$. But it turns out that any abelian subalgebra (that is, having zero brackets) of the correct dimension ℓ is an $L(T)$. This action of $L(T)$ on $L(G)$ determines the roots. In particular, $PSU(n)$ and $SU(n)$ have the same diagram: they have isomorphic Lie algebras and are *locally* isomorphic. Also, the Dynkin diagram does not distinguish any abelian factors of G, since, as we have we observed before, a torus has the empty diagram. The process of going from the Lie group to Dynkin diagram is summarised in the following diagram.

$$\begin{array}{ccc} \text{Reductive Lie algebras over } \mathbf{R} & \xrightarrow{\ \otimes_{\mathbf{R}}\mathbf{C}\ } & \text{Reductive Lie algebras over } \mathbf{C} \\ \uparrow L & & \downarrow \\ \text{Compact connected Lie groups} & \longrightarrow & \text{Dynkin diagrams} \end{array}$$

In the diagram, the bottom arrow is factored round the square. (A lie algebra is *reductive* if its adjoint representation is completely reducible.) The classical classification scheme of Killing and E. Cartan is a classification of simple Lie algebras over $\mathbf{C}$ and gives four infinite families (A_n, B_n, C_n, D_n) and five exceptions (G_2, F_4, E_6, E_7, E_8).

For compact G, the real Lie algebra $L(G)$ is the direct sum of its centre and a semisimple ideal and this ideal is the compact form of its complexification: that is, the Killing form of the ideal is negative definite (see Chapter 6 and [Ha, pp. 220, 193 and 215]). If G, G' are compact Lie groups, then $L(G)$ is isomorphic to $L(G')$ if and only if $L(G) \otimes_{\mathbf{R}} \mathbf{C}$ is isomorphic to $L(G') \otimes_{\mathbf{R}} \mathbf{C}$. Thus the Dynkin diagram determines $L(G)$ and hence, G, locally. In particular, corresponding to each Dynkin diagram, there is a unique compact, connected, simply connected Lie group. To each of the diagrams in the Killing-Cartan classification, there is a unique simple Lie algebra, so a unique connected, simply connected, compact, simple Lie group. The groups $SU(n+1), n \geq 1$, $Sp(n), n \geq 3$ correspond to A_n and C_n, respectively; the groups $Spin(2n+1), n \geq 2$ and $Spin(2n), n \geq 4$, where $Spin(r)$ is the simply connected cover of $SO(r)$ constructed in Chapter 3, correspond to B_n and D_n. All these groups have rank n and are pairwise non-isomorphic.

The purpose of these notes is to study the Lie algebras and simply connected Lie groups corresponding to the non-classical (or "exceptional") Dynkin diagrams (G_2, F_4, E_6, E_7, E_8); these diagrams are listed below (see Chapter 9 for notation and details).

G_2.

F_4.

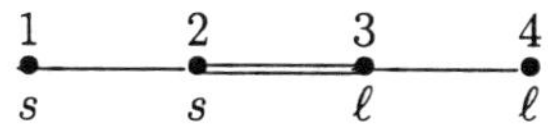

E_6.

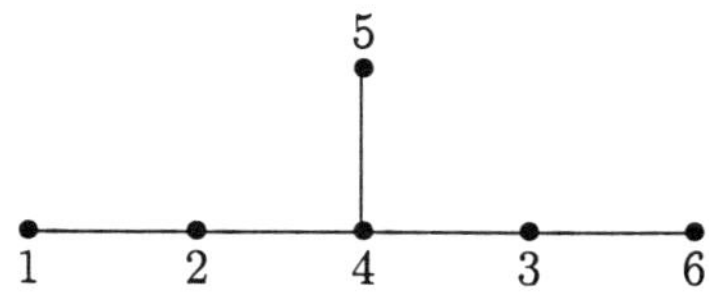

E_7.

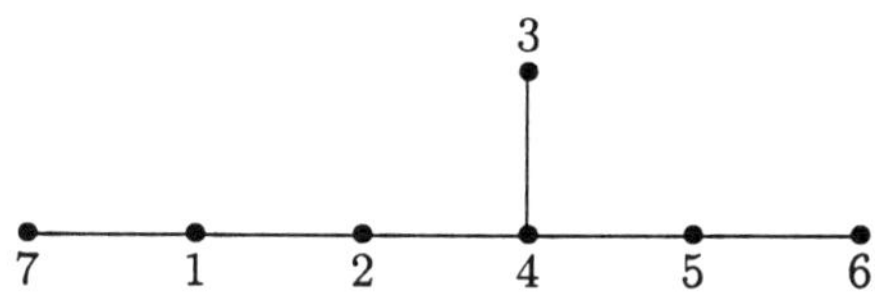

E_8.

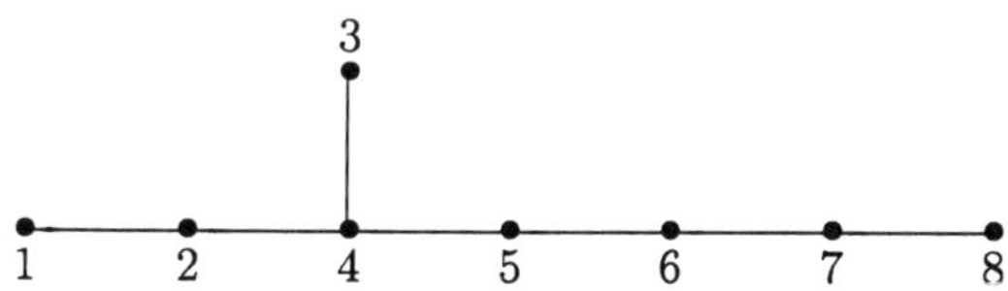

Chapter 2

Clifford algebras

The references for this chapter are [2] and [11, Chapter 3, pp. 240–276].

In studying $Spin(3)$ and $Spin(4)$ we use $\mathbf{H} = \langle i, j \mid i^2 = j^2 = -1, ij = -ji \rangle$, and we could generalise $\mathbf{H}$ by defining a Clifford algebra to be the algebra with generators $e_1, \ldots, e_n$ and relations $e_i^2 = -1$, $e_r e_s = -e_s e_r$, $r \neq s$, but it is better to use a conceptual construction. For this, we begin with V, a non-zero finite dimensional vector space over $\Lambda = \mathbf{R}$ or $\mathbf{C}$.

Definition. $T(V) = \bigoplus_{n \geq 0} \otimes^n V$ *where* $\otimes^0 V = \Lambda$.

We will write $\otimes^n V$ as $V^{\otimes n}$. Define the product in $T(V)$ as follows: if $v_1 \otimes \cdots \otimes v_p \in V^{\otimes p}$ and $w_1 \otimes \cdots \otimes w_q \in V^{\otimes q}$ then their product is $v_1 \otimes \cdots \otimes v_p \otimes w_1 \otimes \cdots \otimes w_q \in V^{\otimes (p+q)}$. For example, if V has a basis $\{x, y\}$, then $T(V)$ has a basis $\{1, x, y, xy, yx, y^2, x^2, \ldots\}$. (Here $xy = x \otimes y$.) In general, $T(V)$ is a free associative algebra, called the *tensor algebra* of V and characterised by the properties of Proposition 2.1.

Proposition 2.1. *There is an inclusion $V \hookrightarrow T(V)$ and for any Λ-linear map $j : V \to A$ of V into an associative Λ-algebra A, there is a unique homomorphism $\theta : T(V) \to A$ extending j.*

$$\begin{array}{ccc} V & \hookrightarrow & T(V) \\ & {\scriptstyle j}\searrow & \vdots\, \theta \\ & & A \end{array}$$

□

Suppose now we give V a symmetric bilinear form $(\ ,\ )$. (For example, on $V = \Lambda^n$, set $(u, v) = -u^T v$.) Let $J = \langle v \otimes v - (v, v) \cdot 1 \mid v \in V \rangle$, a two sided ideal in $T(V)$ and put

$$Cl(V) = T(V)/J,$$

the *Clifford algebra* over V with bilinear form $(\ ,\ )$. Write $i : V \to Cl(V)$ for the composition $V \hookrightarrow T(V) \to T(V)/J = Cl(V)$. In practice we often omit the notation for i and write v for an element of V or its image in $Cl(V)$.

Proposition 2.2. (Universal property of $Cl(V)$.) *For any associative Λ-algebra A and map $j : V \to A$ such that $(jv)^2 = (v,v) \cdot 1 \in A$, there exists a unique map $\theta : Cl(V) \to A$ of Λ-algebras such that $\theta i = j$.*

$$\begin{array}{ccc} V & \xrightarrow{\ i\ } & Cl(V) \\ & {\scriptstyle j}\searrow \quad \swarrow{\scriptstyle \theta} & \\ & A & \end{array}$$

Proof. Immediate. □

$Cl(V)$ is a $\mathbf{Z}_2$-graded algebra and $T(V)$ a $\mathbf{Z}^+$-graded algebra. We put

$$T(V)_0 = \bigoplus_{n\,\text{even}} V^{\otimes n}, \quad T(V)_1 = \bigoplus_{n\,\text{odd}} V^{\otimes n}.$$

Then $v \otimes v - (v,v) \cdot 1 \in T(V)_0$ so that $J = J_0 \oplus J_1$, where $J_i = J \cap T(V)_i$ and

$$Cl(V) = Cl(V)_0 \oplus Cl(V)_1, \quad \text{where} \quad Cl(V)_i = T(V)_i / J_i.$$

If A, B are graded algebras, we put $(A \otimes B)_n = \bigoplus_{p+q=n} A_p \otimes B_q$ and $(a \otimes b)(a' \otimes b') = (-1)^{qr} aa' \otimes bb'$ where $a \in A_p$, $b \in B_q$, $a' \in A_r$, $b' \in B_s$. Then $A \otimes B$ becomes a graded algebra and we have associativity: $(A \otimes B) \otimes C \cong A \otimes (B \otimes C)$.

Proposition 2.3. *If $V = V' \oplus V''$ with $(v', v'') = 0$ for all $v' \in V'$, $v'' \in V''$, then*

$$Cl(V) \cong Cl(V') \otimes Cl(V'').$$

Proof. We check that the right hand side has the universal property which characterises the left hand side. Suppose given $j : V \to A$ such that $(jv)^2 = (v,v) \cdot 1$, then $V' \hookrightarrow V \xrightarrow{j} A$ induces $\theta' : Cl(V') \to A$ and $V'' \hookrightarrow V \xrightarrow{j} A$ induces $\theta'' : Cl(V'') \to A$. We now have to determine the relation between θ' and θ''.

Step 1. If $v' \in V'$, $v'' \in V''$, then $(\theta' v')(\theta'' v'') = -(\theta'' v'')(\theta' v')$ in A.

Proof. $(v' + v'', v' + v'') = (v', v') + (v'', v'')$, so $j(v' + v'') \cdot j(v' + v'') = j(v') \cdot j(v') + j(v'') \cdot j(v'')$ and on expanding we have $j(v')j(v'') + j(v'')j(v') = 0$. □

Step 2. If $v' \in V'$ and $v''_1, \ldots, v''_q \in V''$, then

$$(\theta' v')(\theta''(\textstyle\prod v''_i)) = (-1)^q (\theta''(\textstyle\prod v''_i))(\theta' v').$$

Proof. We can expand $\theta''(\prod v_i'')$ as θ'' is a homomorphism. □

Step 3. If $v_1', \ldots, v_p' \in V'$, $v_1'', \ldots, v_q'' \in V''$, then

$$(\theta'(\Pi v_i'))(\theta''(\Pi v_j'')) = (-1)^{pq}(\theta''(\Pi v_j''))(\theta'(\Pi v_i')).$$

Now define $\theta : Cl(V') \otimes Cl(V'') \to A$ by $x \otimes y \mapsto (\theta' x)(\theta'' y)$. This is a map of algebras by the definition of the product. The required isomorphism comes by taking $A = Cl(V)$. The inclusion $V' \hookrightarrow V$ induces $\theta' : Cl(V') \to Cl(V)$ and $V'' \hookrightarrow V$ induces $\theta'' : Cl(V'') \to Cl(V)$. Now take $\theta : Cl(V') \otimes Cl(V'') \to Cl(V)$ to be $\theta(x \otimes y) = \theta'(x) \otimes \theta''(y)$. □

Remark. Suppose V is 1-dimensional with basis $\{e\}$. Then $Cl(V)$ has basis $\{1, e\}$, because $T(V)$ has basis $\{1, e^1, \ldots\}$ and J has basis $\{e^2 - (e,e)1, e^3 - (e,e)e, \ldots\}$.

Corollary 2.4. *If $\dim_\Lambda V = n$ and $\{e_1, \ldots, e_n\}$ is an orthogonal basis for V with $(e_i, e_j) = \lambda_j \delta_{ij}$, then $\dim_\Lambda Cl(V) = 2^n$ and $\{\prod e_j^{i_j}\}$ is a basis where i_j is 0 or 1.* □

Corollary 2.5. *The maps $\Lambda \hookrightarrow Cl(V)$ and $V \hookrightarrow Cl(V)$ are mono.*

Proof. Among the basis elements are $1, e_1, \ldots, e_n$. □

Corollary 2.6. *Again assume that $\{e_i\}$ is a basis for V and $(e_i, e_j) = -\delta_{ij}$. Then the products in $Cl(V)$ are determined by the following relations: $e_r^2 = -1$, $e_r e_s = -e_s e_r$, $r \neq s$.*

Proof. From the definition, we have $e_r^2 = -1$ and $e_r e_s = -e_r e_s$ from step 1 above. From these relations, we have

$$(\Pi e_k^{i_k})(\Pi e_k^{j_k}) = (-1)^\nu \Pi e_k^{\ell_k},$$

where $\ell_k \equiv i_k + j_k \pmod 2$ and $\nu = \sum_{r \leq s} j_r i_s$. □

Structure maps on Clifford algebras.

First there is the automorphism $\alpha : Cl(V) \to Cl(V)$ induced by $-1 : V \to V$ (minus the identity). Thus $\alpha | Cl(V)_0 = +1$, $\alpha | Cl(V)_1 = -1$.

Example. $Cl(\mathbf{R})$ has generators $1, i$, so $\alpha(1) = 1$, $\alpha(i) = -i$.
$Cl(\mathbf{C})$ has generators $1, i, j, ij$ so $\alpha(1) = 1$, $\alpha(i) = -i$, $\alpha(j) = -j$ and $\alpha(ij) = ij$.

Lemma 2.7. *If $(\ ,\ )$ on V is non-singular and $x \in Cl(V)$ is such that $xv = v(\alpha x)$ for all $v \in V$, then x is a scalar.*

Proof. Over $\Lambda = \mathbf{R}$ or $\mathbf{C}$, we can diagonalise $(\ ,\ )$ and choose a basis $e_1, \ldots, e_n$ in V such that $(e_r, e_s) = \delta_{rs}\lambda_r$, $\lambda_r \neq 0$. Now we can write $x = \sum_I \lambda_I \prod_j e_j^{i_j}$, $\lambda_I \in \Lambda$. Note that e_s is invertible as $e_s e_s = \lambda_s \neq 0$, so that if $xe_s = e_s(\alpha x)$, we have $e_s^{-1} x e_s = \alpha x$. But

$$e_s^{-1}(\prod_j e_j^{i_j})e_s = \begin{cases} (-1)^{\sum i_j} \prod e_j^{i_j} & i_s = 0 \\ (-1)^{-1+\sum i_j} \prod_j e_j^{i_j} & i_s = 1 \end{cases}$$

whilst $\alpha(\prod_j e_j^{i_j}) = (-1)^{\sum i_j} \prod e_j^{i_j}$ in all cases. Thus $xe_s = e_s(\alpha x)$ if and only if $\lambda_I = 0$ whenever $i_s = 1$. Thus $xe_s = e_s(\alpha x)$ for all s if and only if $\lambda_I = 0$ whenever $i_s = 1$ for any s. That is, $\lambda_I \neq 0$ only for $I = (0, \ldots, 0)$ and x is a scalar. □

Remark. If $(\ ,\ ) = 0$, then $Cl(V) \cong \Lambda(V) =$ exterior algebra of V and clearly $x \cdot v = v(\alpha x)$ for all $v \in V$ and all x in this case.

Definition. *Define* $\beta : T(V) \to T(V)$ *by* $\beta(v_1 \otimes \cdots \otimes v_n) = v_n \otimes \cdots \otimes v_1$. *This is an anti-automorphism* $(\beta(xy) = \beta(y)\beta(x))$ *and induces* $\beta : Cl(V) \to Cl(V)$ *with* $\beta|V = 1$. *Also* $\gamma = \alpha\beta = \beta\alpha : Cl(V) \to Cl(V)$ *is an anti-automorphism such that* $\gamma|V = -1$.

Example 1. $Cl(\mathbf{R})$ is generated by $1, i$, and $\beta(1) = 1$, $\beta(i) = i$, $\gamma(1) = 1$, $\gamma(i) = -i$.

Example 2. The algebra $\mathbf{H} = Cl(\mathbf{C})$ has basis $1, i, j, ij = k$, and the action of α, β, γ is given by

	1	i	j	k
α	1	$-i$	$-j$	k
β	1	i	j	$-k$
γ	1	$-i$	$-j$	$-k$

Note that for $Cl(\mathbf{R}) = \mathbf{C}$ or $Cl(\mathbf{C}) = \mathbf{H}$, γ is the usual conjugation.

Chapter 3

The Spin groups

The references for this chapter are also [2] and [11]. In addition see [Br].

Let $V = \Lambda^n$, $\Lambda = \mathbf{R}$ or $\mathbf{C}$ and $(\ ,\ )$ be the inner product $(u, v) = u^T v$ and form $Cl(V)$ with respect to $-(\ ,\ )$. Let $e_1, \ldots, e_n$ be the standard basis in V, so that in $Cl(V)$ we have the products $e_r e_s$ $(r < s$; there are $n(n-1)/2$ of them) and these span a Lie algebra with $[e_r e_s, e_t e_u] = e_r e_s e_t e_u - e_t e_u e_r e_s$. Thus if r, s, t, u are different, $[e_r e_s, e_t e_u] = 0$ since $e_s e_t e_u = -e_t e_s e_u = e_t e_u e_s$ and $e_r e_s e_t e_u = e_r e_t e_u e_s = e_t e_u e_r e_s$. If r, t, u are different, $[e_r e_u, e_t e_u] = 2 e_r e_t$ since $e_r e_u e_t e_u = -e_r e_t e_u e_u = e_r e_t$ and $e_t e_u e_r e_u = e_t e_r = -e_r e_t$. Finally, $[e_r e_s, e_r e_s] = 0$. We wish to see this Lie algebra as the Lie algebra of a Lie group.

Definition. *$Pin(V) \subset Cl(V)$ is the subset of elements x such that*

(i) $x(\gamma x) = (\gamma x)x = 1$,

(ii) *the map $\pi x : V \to Cl(V)$ defined by $(\pi x)v = xv(\beta x)$ maps $V \subset Cl(V)$ into V.*

Example. $Pin(\mathbf{R}) = \{ z \in Cl(\mathbf{R}) \mid z\bar{z} = 1,\ z^2 \in \mathbf{R} \} = \{\pm 1, \pm i\}$.

Proposition 3.1. (i) *$Pin(V)$ is a subgroup of the invertible elements of $Cl(V)$ and its Lie algebra is that given above (i.e. spanned by $\{ e_r e_s \mid r < s \}$).*

(ii) *The map $\pi : Pin(V) \to O(V)$, where $O(V)$ is the group of Λ-linear maps $V \to V$ preserving $-(\ ,\ )$, is an epimorphism with $\ker \pi = \{\pm 1\}$.*

The closed subsets $\pi^{-1}(\det^{-1}(1))$, $\pi^{-1}(\det^{-1}(-1))$ of $Pin(V)$ are in $Cl(V)_0$, $Cl(V)_1$ respectively and are connected for $n \geq 2$.

Explanations. We treat the real and complex cases simultaneously. If $\Lambda = \mathbf{R}$, $O(V) \cong O(n)$ and if $\Lambda = \mathbf{C}$, $O(V)$ is the complex orthogonal group, the subgroup of $GL(n, \mathbf{C})$ such that $\sum z_r^2$ is preserved. Linear maps $f : V \to V$ in either of these groups must have determinant $\det f = \pm 1$, so $O(V) = \det^{-1}(1) \cup \det^{-1}(-1)$. If $\pi^{-1}(\det^{-1}(1))$, $\pi^{-1}(\det^{-1}(-1))$ are connected, $\det^{-1}(1)$, $\det^{-1}(-1)$ must also be connected.

(Let us recall how to show that $\det^{-1}(1)$ $(= SO(V))$ and $\det^{-1}(-1)$ are connected. It suffices to prove that $SO(V)$ is connected by induction on $n = \dim V$. Clearly $SO(V)$ is connected for $n = 1$, so assume that $SO(V)$ is connected for $\dim V = n-1$, $n > 1$. By applying [S, p. 377, Corollary 13] to $SO(n) \to SO(n)/SO(n-1)$, we see it is sufficient to show that $SO(n)/SO(n-1)$ is connected and this may be identified with $\{ v \in V \mid (v,v) = 1 \}$, $n \geq 2$, under the map $[A] \mapsto Ae_n$ where $e_n^T = (0,0,\dots,0,1)$. For $\Lambda = \mathbf{R}$, $SO(n)/SO(n-1)$ is identified with $S^{n-1} \subset \mathbf{R}^n$ and this is connected $(n \geq 2)$ and for $\Lambda = \mathbf{C}$, it is identified with $\{ z \in \mathbf{C}^n \mid z_1^2 + \cdots + z_n^2 = 1 \}$. Put $z = x + iy$, $x, y \in \mathbf{R}^n$ and consider the locus $Re(z_1^2 + \cdots + z_n^2) = 1$, that is $x_1^2 + \cdots + x_n^2 = y_1^2 + \cdots + y_n^2 + 1$. This is homeomorphic to $S^{n-1} \times \mathbf{R}^n$, hence connected. On $Re(z) = 1$, we can make $\sqrt{z}$ into a well defined continuous function and by mapping $(z_1, \dots, z_n)$ to $\left(1/\sqrt{z_1^2 + \cdots + z_n^2}\right)(z_1, \dots, z_n)$, we get a retraction from the locus $Re(z_1^2 + \cdots + z_n^2) = 1$ to the locus $z_1^2 + \cdots + z_n^2 = 1$, so the latter is connected as well.) □

Finally, $Spin(V)$ is the subgroup $\pi^{-1}(\det^{-1}(1)) = Pin(V) \cap Cl(V)_0$. It comes with a homomorphism $\pi : Spin(V) \to SO(V)$.

Proof (of Proposition 3.1). It is straightforward to check that $Pin(V)$ is a closed subgroup of invertible elements of $Cl(V)$. Next we show that $\pi(x) \in O(V)$:

$$\begin{aligned} ((\pi x)v, (\pi x)v) &= ((\pi x)v)^2 = -xv(\beta x)(\alpha x)v(\gamma x) \quad (\text{as } -(\pi x)v = -(\alpha x)v(\gamma x)) \\ &= -xvv\gamma(x) = (v,v)x\gamma x = (v,v). \end{aligned}$$

This checks that $\pi(x) \in O(V)$. It is easy to see that π is a homomorphism.

Suppose now that $x \in \ker \pi$. Then $v = xv\beta x$, for all $x \in V$, so $v\alpha x = xv(\beta x)(\alpha x) = xv$. By Lemma 2.7, x is a scalar. But $x\gamma x = 1$. Thus $x^2 = 1$, $x = \pm 1$ and $\ker \pi \subset \{\pm 1\}$. Certainly $\pm 1 \in \ker \pi$, so $\ker \pi = \{\pm 1\}$. Next we need to establish the following

Claim. *$L(\pi) : L(Pin(V) \cap Cl(V)_0) \longrightarrow L(O(V))$ is epi.*

Proof (of Claim). One checks that $Pin(V) \cap Cl(V)_0$ is again a closed subgroup. For $r < s$, $(e_r e_s)(e_r e_s) = -e_r e_r e_s e_s = -1$ so $x = e^{(e_r e_s)t} = \cos t + (\sin t) e_r e_s$ is defined for $\mathbf{R}$ or $\mathbf{C}$. Now,

$$\pi(x) = \begin{array}{c} \\ \\ \\ r \\ \\ \\ \\ s \\ \\ \\ \\ \end{array} \overset{\qquad\qquad\qquad r \qquad\qquad\qquad\qquad\qquad s}{\begin{pmatrix} 1 &&&&&&&&& \\ & \ddots &&&&&&&& \\ && 1 &&&&&&& \\ &&& \cos 2t &&&& -\sin 2t && \\ &&&& 1 &&&&& \\ &&&&& \ddots &&&& \\ &&&&&& 1 &&& \\ &&& \sin 2t &&&& \cos 2t && \\ &&&&&&&& 1 & \\ &&&&&&&&& \ddots \\ &&&&&&&&&& 1 \end{pmatrix}}$$

since

$$\Big(\cos t + (\sin t)e_r e_s\Big)e_u\Big(\cos t - (\sin t)e_r e_s\Big) = \begin{cases} e_u & \text{if } u \neq r, s \\ \Big(\cos 2t + (\sin 2t)e_r e_s\Big)e_u & \text{if } u = r, s. \end{cases}$$

Thus $\pi(x)$ maps V to V, so $x \in Pin(V)$. In fact $x \in Pin(V) \cap Cl(V)_0$. We see that the Λ-multiples of $e_r e_s$ lie in $L(Pin(V) \cap Cl(V)_0)$ and map under $L(\pi)$ to the Λ-multiples of $\begin{pmatrix} 0 &&&& \\ & \ddots && -2 & \\ && 0 && \\ & 2 && \ddots & \\ &&&& 0 \end{pmatrix} = (\pi(x))'|_{t=0}$, which form a Λ-basis for $L(O(V))$, the skew-symmetric matrices. This proves that $L(\pi)$ is epi. $\square$

It follows from the Claim that $L(\pi) : L(Pin(V)) \to L(O(V))$ is also epi; it must be mono as $\ker \pi$ is finite. Thus

$$L(Pin(V) \cap Cl(V)_0) = L(Pin(V)),$$

and so $L(\pi) : L(Spin(V)) \to L(O(V))$ is an isomorphism. Since $L(\pi)$ maps the Λ-span of $\{ e_r e_s \mid r < s \}$ onto $L(O(V))$ and $L(\pi)$ is an isomorphism, $\{ e_r e_s \mid r < s \}$ must be a Λ-basis for $L(Pin(V))$.

Now, using exp and log, we find that π maps a small neighbourhood of 1 in $Pin(V) \cap Cl(V)_0$ onto a small neighbourhood of 1 in $O(V)$. Thus π maps the identity component of $Pin(V) \cap Cl(V)_0$ onto the identity component of $O(V)$, i.e. onto $SO(V)$. But $\{\pm 1\}$ is contained in the identity component of $Pin(V) \cap Cl(V)_0$ if $n \geq 2$, since $\cos t + (\sin t)e_1 e_2$, $t \in [0, \pi]$, is a path from 1 to -1 in $Pin(V) \cap Cl(V)_0$. So the identity component of $Pin(V)_0 \cap Cl(V)_0 = \pi^{-1}(SO(V))$ in $Pin(V)$. Thus $\pi^{-1}(\det^{-1}(1))$ is connected and contained in

$Cl(V)_0$. To show that $\pi^{-1}(\det^{-1}(-1))$ is connected and complete the proof, it is sufficient to give an element in $\pi^{-1}(\det^{-1}(-1))$ so that multiplying by it will send $\pi^{-1}(\det^{-1}(1))$ to $\pi^{-1}(\det^{-1}(-1))$. The element e_1 will do, for one checks that it lies in $Pin(V)$ and covers the reflection $diag(-1, 1, \ldots, 1)$. □

Remark 1. Over $\mathbf{R}$, the maximal torus $\widetilde{T}$ in $Spin(m)$, $m = 2n$ or $2n+1$, is usually taken to consist of the elements $\prod_{r=1}^{n}\left(\cos\frac{1}{2}x_r + (\sin\frac{1}{2}x_r)e_{2r-1}e_{2r}\right)$ in $Cl(V)$, $x_r \in \mathbf{R}$, which corresponds to

$$\begin{pmatrix} \cos x_1 & -\sin x_1 & & & & & & 0 & \\ \sin x_1 & \cos x_1 & & & & & & & \\ & & \cos x_2 & -\sin x_2 & & & & & \\ & & \sin x_2 & \cos x_2 & & & & & \\ & & & & \ddots & & & & \\ & 0 & & & & \cos x_n & -\sin x_n & & \\ & & & & & \sin x_n & \cos x_n & & \\ & & & & & & & & 1 \end{pmatrix}$$

where row $n+1$ has 1 as the last entry if $m = 2n+1$, and is empty for $m = 2n$.

Remark 2. For $\Lambda = \mathbf{R}$, we can show directly that every element in $Spin(m)$ is conjugate to an element in $\widetilde{T}$ from the corresponding result for $SO(m)$, as follows. Take $g \in Spin(m)$, $\pi g \in SO(m)$. Then there exists $h \in SO(m)$ such that $h(\pi g)h^{-1} \in T \subset SO(m)$. Lift h to $h' \in Spin(m)$ under $\pi : Spin(m) \to SO(m)$ to get $h'gh'^{-1} \in \widetilde{T}$.

(Alternative proof. For $\pi g \in SO(m)$, take a basis $b_1, \ldots, b_m$ of $\mathbf{R}^m$ with respect to which πg has matrix as in Remark 1 above. Then $\prod_{r=1}^{n}\left(\cos\frac{1}{2}x_r + (\sin\frac{1}{2}x_r)b_{2r-1}b_{2r}\right) \in Spin(m)$ covers g.)

Chapter 4

Clifford modules and representations

The references for this chapter are [8], [11] and [2].

Since $Spin(V) \subset Cl(V)_0$, any $Cl(V)_0$-module is a representation of $Spin(V)$, and some important representations of $Spin(n)$ arise in this way.

Proposition 4.1. *The algebras $Cl(V)$ and $Cl(V)_0$ are semi-simple, so all their representations are completely reducible.*

Proof. Let $\{e_1, \ldots, e_m\}$ be the standard basis in $V = \Lambda^m$ and consider $E = \{\pm \prod_{j=1}^{m} e_j^{i_j} \mid i_j = 0 \text{ or } 1\}$, a subgroup of order 2^{m+1} of $Cl(V)$, corresponding to the matrices $diag(\pm 1, \ldots, \pm 1)$ of $O(m)$. In $Cl(V)_0$ consider the subgroup E_0 of 2^m elements with $\sum_j i_j$ even, $E_0 \subset E \subset Pin(V)$. To avoid confusion, write ν for $-1 \in Cl(V)_0$ when considered an element of E_0, E or $Pin(V)$; ν is the generator of $\ker \pi$. A module over $Cl(V)$ gives a representation of E in which ν acts as -1. Conversely, a representation of E in which ν acts as -1 gives a module over $\Lambda(E)/\langle \nu + 1 \rangle = Cl(V)$, where $\Lambda(E)$ is the group ring over Λ of E and Λ is **R** or **C**. The representation theory of $Cl(V)$ can thus be inferred from that of the finite group E and in particular all representations are completely reducible. We argue similarly for $Cl(V)_0$, replacing E by E_0. □

Proposition 4.2. *If $\dim V$ is odd, $m = 2n+1$ say, then $Cl(V)_0$ has one irreducible representation Δ of degree 2^n affording a representation Δ of $Spin(2n+1)$ with weights $\frac{1}{2}(\pm x_1 \pm x_2 \pm \cdots \pm x_n)$, so there are 2^n of these weights. If $\dim V = m = 2n$, then $Cl(V)_0$ has two irreducible representations Δ^+, Δ^- of degree 2^{n-1} affording representations Δ^+, Δ^- of $Spin(2n)$ having* *weights*

$$\tfrac{1}{2}(\pm x_1 \pm x_2 \pm \cdots \pm x_n), \quad \textit{(even number of } - \textit{ signs)} \quad \textit{for } \Delta^+, \quad \textit{and}$$
$$\tfrac{1}{2}(\pm x_1 \pm x_2 \pm \cdots \pm x_n), \quad \textit{(odd number of } - \textit{ signs)} \quad \textit{for } \Delta^-.$$

There are 2^{n-1} *weights. If* $\Lambda = \mathbf{C}$, *these are complex-analytic representations of* $Spin_{\mathbf{C}}(m)$.

Proof. By Schur's Lemma, ν acts on any irreducible representation as either 1 or as -1. The ones in which it acts as 1 are representations of $E_0/\langle\nu\rangle$ which is an abelian group of order 2^{m-1}, so there are exactly 2^{m-1} 1-dimensional representations of E_0 in which ν acts as 1. Since the kernel of $E_0 \to E_0/\langle\nu\rangle$ has exactly two elements, the conjugacy classes in E_0 are either one element (if the element is central) or two elements $\pm g$.

Lemma. *The centre of* E_0 *is* $\{\pm 1\}$ *if* $m = 2n+1$ *and* $\{\pm 1, \pm \prod_1^{2n} e_i\}$ *if* $m = 2n$.

Proof. If we conjugate $g = \prod_1^m e_j^{i_j}$ with $e_r e_s$ where $i_r = 1, i_s = 0$, we change its sign; so if g is in the centre, $g = \pm 1$ or $\pm e_1 e_2 \ldots e_m$. The latter is in the centre only for m even. □

Recalling that the isomorphism classes of irreducible representations (over $\mathbf{C}$) of a finite group are in 1-1 correspondence with the conjugacy classes, we see that E_0 has one (respectively two) more irreducible class(es) of representation(s) than $E_0/\langle\nu\rangle$ if $m = 2n+1$ (respectively, $m = 2n$). Let $F \subset E_0$ be the subgroup generated by $e_1e_2, \ldots, e_{2r-1}e_{2r}, \ldots, e_{2n-1}e_{2n}$. This is an abelian group of order 2^{n+1}, so $|E_0 : F| = 2^n$ if $m = 2n+1$ and $= 2^{n-1}$ if $m = 2n$.

Choose a complex 1-dimensional representation W of F in which ν acts as -1 and $e_{2r-1}e_{2r}$ acts as $i\,\varepsilon_r$, $\varepsilon_r = \pm 1$, $i = \sqrt{-1}$. Form

$$\mathbf{C}(E_0) \otimes_{\mathbf{C}(F)} W \qquad \text{(i.e. the induced representation, } Ind_F^{E_0} W\text{)},$$

where $\mathbf{C}(E_0)$ is the complex group algebra of E_0. This is now a representation of E_0 with degree 2^n for $m = 2n+1$ and 2^{n-1} for $m = 2n$ with ν acting as -1. It has a basis consisting of

$$\prod_{i\ \mathrm{odd}=1}^{2n+1} e_i^{j_i} \quad \text{with} \quad \sum_1^{2n+1} j_i \ \text{even}, \quad \text{if } m = 2n+1,$$
$$\prod_{i\ \mathrm{odd}=1}^{2n-1} e_i^{j_i} \quad \text{with} \quad \sum_1^{2n-1} j_i \ \text{even}, \quad \text{if } m = 2n.$$

When $m = 2n+1$, there are 2^n choices for W (because we can choose n signs ε_r), E_0/F permutes them transitively and, by conjugating with $e_{2r}e_{2r+1}$, we can change the sign of ε_r without changing anything else. Each appears once in $\mathbf{C}(E_0)\otimes_{\mathbf{C}(F)} W$. We thus get a representation, Δ of E_0 with character 2^n at 1, -2^n at -1, and 0 elsewhere. By the orthogonality relations, Δ is an irreducible representation of E_0. When $m = 2n$, there are 2^n choices for W and under E_0/F they fall into two orbits: those with $\prod \varepsilon_r = 1$ and those with $\prod \varepsilon_r = -1$. We can only change the sign of an even number of ε_r: conjugating with $e_{2r}e_{2s}$, $r < s$, changes the sign of both ε_r and ε_s. Hence we have one irreducible representation of E_0 which as a representation of F contains all the W with $\varepsilon = \prod \varepsilon_r = 1$ and another containing all the W with $\varepsilon = \prod \varepsilon_r = -1$. The character of these representations is 2^{n-1} at 1, -2^{n-1} at -1, $i^n\varepsilon$ at $\prod_1^{2n} e_i$, $-i^n\varepsilon$ at $-\prod_1^{2n} e_i$, and zero elsewhere. This is an irreducible character by the orthogonality relations. Thus these are inequivalent representations. □

Remarks 1. These representations are self dual except in the case $m \equiv 2 \bmod 4$ when the two representations are dual to each other, see Proposition 4.3 below. Here dual means complex conjugate.

2. If we take $\Lambda = \mathbf{C}$ and think of the complex form of *Spin*(V), we get complex analytic representations. But to give the weights it is sufficient to take $\Lambda = \mathbf{R}$ and consider the compact form of the group. This is what is usually meant when we write *Spin*(V).

3. We calculate the weights of Δ for $\Lambda = \mathbf{R}$ as follows: The element

$$\prod_{r=1}^{n}\Big(\cos(2\pi)(\tfrac{1}{2}x_r) + (\sin(2\pi)(\tfrac{1}{2}x_r))e_{2r-1}e_{2r}\Big)$$

of the maximal torus acts on Δ with eigenvalues

$$\prod_{r=1}^{n}\Big(\cos(2\pi)(\tfrac{1}{2}x_r) + (\sin(2\pi)(\tfrac{1}{2}x_r))(\varepsilon_r i)\Big) = \exp(2\pi i \sum_{r=1}^{n} \tfrac{1}{2}\varepsilon_r x_r)$$

and weights $\frac{1}{2}\sum_{r=1}^{n} \varepsilon_r x_r$, where $\varepsilon_r = \pm 1$. □

Proposition 4.3. *The representation Δ of Spin$(2n+1)$ is self dual. The representations Δ^+, Δ^- of Spin$(2n)$ are self dual if n is even and dual to each other if n is odd.*

Proof. We have to prove the isomorphism in various senses of M^* with N where M, N are from Δ, $\Delta^\pm$ as stated. Well, E_0 acts on $M^* = Hom_{\mathbf{C}}(M,\mathbf{C})$ as $(gh)(m) = h(g^{-1}m)$, $m \in M$, and $Cl(\mathbf{C}^m)_0 \cong \mathbf{C}(E_0)/\langle \nu + 1\rangle$ acts on M^* as $(ah)(m) = h((\gamma a)m)$ where $\gamma : Cl(\mathbf{C}^m)_0 \to Cl(\mathbf{C}^m)_0$ is the anti-automorphism of Chapter 2. From the proof of Proposition 4.2, we have an isomorphism of the representations N and M^* of the finite group E_0, which is an isomorphism of $Cl(\mathbf{C}^m)_0$-modules. But $Spin(m) \subset Cl(\mathbf{C}^m)_0$, so the isomorphism preserves the action of elements of $Spin(m)$ provided this action is defined by $(gh)m = h(g^{-1}m)$, which is the usual action. Note that this argument gives an isomorphism of the representations of the complex Spin groups if we want. □

For the next result note that the inclusion $\Lambda^m \hookrightarrow \Lambda^{m+1}$ induces an inclusion $Cl(\Lambda^m) \hookrightarrow Cl(\Lambda^{m+1})$ so we get $Spin(m) \hookrightarrow Spin(m+1)$ covering the usual map $SO(m) \hookrightarrow SO(m+1)$, where $A \mapsto \left(\begin{smallmatrix} A & 0 \\ 0 & 1 \end{smallmatrix}\right)$.

Proposition 4.4. *Under the inclusions*

$$Spin(2n) \hookrightarrow Spin(2n+1) \hookrightarrow Spin(2n+2),$$

we have

$$\Delta^+ + \Delta^- \longleftarrow \Delta \begin{array}{l} \nearrow \Delta^+ \\ \searrow \Delta^- \end{array} .$$

We will prove this proposition with the next one. We have $Cl(\Lambda^p) \otimes_\Lambda Cl(\Lambda^q) \cong Cl(\Lambda^{p+q})$ which gives $Spin(p) \times Spin(q) \to Spin(p+q)$ and $E_0(p) \times E_0(q) \to E_0(p+q)$. These cover the map $SO(p) \times SO(q) \hookrightarrow SO(p+q)$ where $(A, B) \mapsto \left(\begin{smallmatrix} A & 0 \\ 0 & B \end{smallmatrix}\right)$. Here the last map is an injection but the previous two are not : they carry (ν, ν) to 1.

Proposition 4.5.

(i) *Under*
$$\begin{array}{ccc} Spin(2r) \times Spin(2s) & \longrightarrow & Spin(2r+2s) \\ \Delta^+ \otimes \Delta^+ + \Delta^- \otimes \Delta^- & \longleftarrow & \Delta^+ \\ \Delta^+ \otimes \Delta^- + \Delta^- \otimes \Delta^+ & \longleftarrow & \Delta^-. \end{array}$$

(ii) *Under*
$$\begin{array}{ccc} Spin(2r) \times Spin(2s+1) & \longrightarrow & Spin(2r+2s+1) \\ \Delta^+ \otimes \Delta + \Delta^- \otimes \Delta & \longleftarrow & \Delta. \end{array}$$

(iii) *Under*
$$\begin{array}{ccc} Spin(2r+1) \times Spin(2s+1) & \longrightarrow & Spin(2r+2s+2) \\ \Delta \otimes \Delta & \longleftarrow & \Delta^+ \\ \Delta \otimes \Delta & \longleftarrow & \Delta^-. \end{array}$$

Proof. The proofs of the last two properties are closely similar. We have two modules M, N to be proved isomorphic in various senses, namely the module on the left and the pull back of the module on the right. In each case the weights agree, so the characters agree on the compact group $Spin_{\mathbf{R}}(p) \times Spin_{\mathbf{R}}(q)$.

For example, in the first case, if we take a weight $\frac{1}{2}(\varepsilon_1 x_1 + \cdots + \varepsilon_{r+s} x_{r+s})$, with $\varepsilon_1 \varepsilon_2 \ldots \varepsilon_{r+s} = 1$, then we get on the right $\frac{1}{2}(\varepsilon_1 x_1 + \cdots + \varepsilon_r x_r) + \frac{1}{2}(\varepsilon_{r+1} x_{r+1} + \cdots + \varepsilon_{r+s} x_{r+s})$ either with $\varepsilon_1 \varepsilon_2 \ldots \varepsilon_r = 1$ and $\varepsilon_{r+1} \ldots \varepsilon_{r+s} = 1$ or with $\varepsilon_1 \varepsilon_2 \ldots \varepsilon_r =$

-1 and $\varepsilon_{r+1}\cdots\varepsilon_{r+s} = -1$. This shows that the characters agree on the group $Spin_{\mathbf{R}}(p) \times Spin_{\mathbf{R}}(q)$. Hence the characters agree on the finite groups $E_0(p) \times E_0(q)$ so the modules are isomorphic as representations of these finite groups and thus isomorphic as modules over $\Lambda(E_0(p) \times E_0(q))$. But they are isomorphic as modules over $\Lambda(E_0(p))/\langle\nu+1\rangle \otimes_\Lambda \Lambda(E_0(q))/\langle\nu+1\rangle = Cl(\Lambda^p)_0 \otimes_\Lambda Cl(\Lambda^q)_0$. So, the modules are isomorphic as representations of $Spin_\Lambda(p) \times Spin_\Lambda(q)$, where $Spin_\Lambda(p) = Spin(\Lambda^p)$ with $\Lambda = \mathbf{R}$ or $\mathbf{C}$. □

This gives several representations of $SO(n+1)$ since $\nu \in Spin(2n+1)$ acts as $(-1)\otimes(-1) = 1$ on $\Delta\otimes\Delta$. We know that $SO(m)$ acts on $\mathbf{R}^m = \lambda^1$ (where λ^i is the i-th exterior power of dim $\binom{m}{i}$) and $SO(2n)$ acts on $\lambda^n = \lambda^{n+} + \lambda^{n-}$. See [1, p. 177] and [12, p. 164]. The next theorem gives $\Delta^+ \otimes \Delta^-$ in terms of these exterior powers.

Theorem 4.6. (i) *As representations of $Spin(2n)$ we have*

$$\begin{aligned}
\sigma^2(\Delta^+) &= (\lambda^n)^+ + \lambda^{n-4} + \lambda^{n-8} + \cdots,\\
\lambda^2(\Delta^+) &= \lambda^{n-2} + \lambda^{n-6} + \cdots,\\
\sigma^2(\Delta^-) &= (\lambda^n)^- + \lambda^{n-4} + \lambda^{n-8} + \cdots,\\
\lambda^2(\Delta^-) &= \lambda^{n-2} + \lambda^{n-6} + \cdots,\\
\Delta^+ \otimes \Delta^- &= \lambda^{n-1} + \lambda^{n-3} + \lambda^{n-5} + \cdots,
\end{aligned}$$

where in each summation we have non-negative exterior powers.

(ii) *As representations of $Spin(2n+1)$ we have*

$$\begin{aligned}
&\sigma^2(\Delta) = \lambda^n + (\lambda^{n-3} + \lambda^{n-4}) + \cdots + (\lambda^{n-4i-3} + \lambda^{n-4i-4}) + \cdots,\\
&\lambda^2(\Delta) = (\lambda^{n-1} + \lambda^{n-2}) + (\lambda^{n-5} + \lambda^{n-6}) + \cdots + (\lambda^{n-4i-1} + \lambda^{n-4i-2}) + \cdots.
\end{aligned}$$

where again in each summation we have non-negative exterior powers.

Note. For a representation V, $\sigma^2(V)$ is the second symmetric power, see [La].

Proof. We will prove just the result for $\Delta^+ \otimes \Delta^-$. First we establish the sum of the last two results, that is for $\Delta\otimes\Delta$, which has $Spin(2n+1)$ acting on it as $g(m\otimes m') = gm \otimes gm'$.

Since Δ is self dual, we have $\Delta\otimes\Delta \cong \Delta\otimes\Delta^* \cong Hom_{\mathbf{C}}(\Delta,\Delta)$; g acts on $h \in Hom_{\mathbf{C}}(\Delta,\Delta)$ to give ghg^{-1}. Since $Cl(V)_0$ is semi-simple and Δ is its only irreducible representation up to isomorphism, we have $Hom_{\mathbf{C}}(\Delta,\Delta) \cong Cl(V)_0$ with g acting on $a \in Cl(V)_0$ by gag^{-1}. This is the same as the action of G we know already: $G = Spin(2n+1)$ acts on V and that induces automorphisms of $Cl(V)_0$. (The proof of this is that since $g \in Pin(V) \cap Cl(V)_0$, we have

$g^{-1} = \gamma g = \beta g$.) But this representation of G on $Cl(V)$ splits as can be seen as follows. We define a map $V^{\otimes i} \to Cl(V)$ by

$$v_1 \otimes v_2 \otimes \cdots \otimes v_i \mapsto \sum_{\sigma \in \Sigma_i} \frac{\varepsilon(\sigma)}{i\,!} v_{\sigma(1)} \otimes v_{\sigma(2)} \otimes \cdots \otimes v_{\sigma(i)}$$

(where Σ_i is the symmetric group) which leads to vector space isomorphisms $\oplus \lambda^i V \to Cl(V)$ and $\oplus \lambda^{2r}(V) \to Cl(V)_0$. Hence $\Delta \otimes \Delta \cong \lambda^0 + \lambda^2 + \lambda^4 + \cdots + \lambda^{2n}$. But $\lambda^i \cong \lambda^j$ when $i + j = 2n + 1$, so $\Delta \otimes \Delta \cong \lambda^0 + \lambda^1 + \cdots + \lambda^n$, which is the sum of $\sigma^2(\Delta)$ and $\lambda^2(\Delta)$.

The same argument works for $Spin(2n)$, except with two irreducible representations Δ^+ and Δ^- we have:

$$Hom_{\mathbf{C}}(\Delta^+, \Delta^+) + Hom_{\mathbf{C}}(\Delta^-, \Delta^-) \cong Cl(\mathbf{C}^{2n})_0 = \lambda^0 + \lambda^2 + \lambda^4 + \cdots + \lambda^{2n}.$$

First consider n odd so that Δ^- and Δ^+ are dual to each other. Then

$$2\Delta^+ \otimes \Delta^- = \lambda^0 + \lambda^2 + \lambda^4 + \cdots + \lambda^{2n} = 2\lambda^{n-1} + 2\lambda^{n-3} + \cdots + 2.$$

When n is even, we have

$$\Delta^+ \otimes \Delta^+ + \Delta^- \otimes \Delta^- = \lambda^n + 2\lambda^{n-2} + 2\lambda^{n-4} + \cdots + 2\lambda^0.$$

On the other hand, we can take the formula for $\Delta \otimes \Delta$ and restrict it from $Spin(2n+1)$ to $Spin(2n)$ and get

$$(\Delta^+ + \Delta^-) \otimes (\Delta^+ + \Delta^-) = \lambda^n + 2\lambda^{n-1} + 2\lambda^{n-2} + \cdots + 2\lambda^0.$$

Subtracting off the formula for $\Delta^+ \otimes \Delta^+ + \Delta^- \otimes \Delta^-$, we have

$$2(\Delta^+ \otimes \Delta^-) = 2\lambda^{n-1} + 2\lambda^{n-3} + \cdots + 2\lambda^1,$$

so

$$\Delta^+ \otimes \Delta^- \quad = \lambda^{n-1} + \lambda^{n-3} + \cdots .$$

□

Corollary 4.7. *There is a bilinear map $\mu : M \otimes M \to \mathbf{C}$ which is invariant, non-zero and*

symmetric in cases	$M = \Delta^+,\ \Delta^-$	*if* $2n \equiv 0 \pmod 8$,
	$M = \Delta$	*if* $2n + 1 \equiv 1$ *or* $7 \pmod 8$;
anti-symmetric in cases	$M = \Delta^+,\ \Delta^-$	*if* $2n \equiv 4 \pmod 8$,
	$M = \Delta$	*if* $2n + 1 \equiv 3$ *or* $5 \pmod 8$.

Proof 1. From Theorem 4.6, we can see where a summand $\Lambda^0 = \mathbf{C}$ arises.

Proof 2. We always have an evaluation map $M^* \otimes M \to \mathbf{C}$. When M is self dual, this gives an invariant non-zero map $M \otimes M \to \mathbf{C}$. When M is $\mathbf{C}$-irreducible, this must be either symmetric or anti-symmetric but not both. Thus we can find out which case holds by testing on one pair of vectors. Since $\mu \neq 0$, there are eigenvectors of the T-action such that $\mu(w \otimes w') \neq 0$. Moreover, w, w' must be eigenvectors of opposite weights, for if $tw = \lambda(t)w$, $tw' = \lambda'(t)w'$, then $\mu(w \otimes w') = \mu(tw \otimes tw') = \mu(\lambda(t)w \otimes \lambda'(t)w') = \lambda(t)\lambda'(t)\mu(w \otimes w')$ and $\lambda(t)\lambda'(t) = 1$ for all $t \in T$.

Any w' of weight opposite to w will do since all are scalar multiples of one another. Let us construct one by taking

$$\begin{aligned} e &= e_1 e_3 e_5 \ldots e_{2n+1} \quad &&(2n+1 \equiv 3 \bmod 4), \quad \text{and} \\ e &= e_1 e_3 e_5 \ldots e_{2n-1} \quad &&\text{in all other cases.} \end{aligned}$$

As $0 \neq \mu(w \otimes ew) = \mu(ew \otimes e^2 w)$, $e^2 w$ must be an eigenvalue for the same weight as w, so we must have $e^2 w = \lambda w$ for some scalar λ. So, $0 \neq \mu(w \otimes ew) = \lambda\mu(ew \otimes w)$ and μ is symmetric if $\lambda = 1$, anti-symmetric if $\lambda = -1$.

$$\text{If } e = e_1 e_3 e_5 \ldots e_{2n-1}, \text{ then } e^2 = (-1)^{\frac{1}{2}n(n+1)} = \begin{cases} 1 & \text{if } 2n \equiv 0 \bmod 8, \\ -1 & \text{if } 2n \equiv 4 \bmod 8. \end{cases}$$

$$\text{If } e = e_1 e_3 e_5 \ldots e_{2n+1}, \text{ then } e^2 = (-1)^{\frac{1}{2}(n+1)(n+2)} = \begin{cases} 1 & \text{if } 2n+1 \equiv 7 \bmod 8, \\ -1 & \text{if } 2n+1 \equiv 3 \bmod 8. \end{cases}$$

□

Corollary 4.8. *The following representations of the compact groups* $Spin_{\mathbf{R}}(m)$ *are*

(*a*) *real*	: Δ^+, Δ^-	*if*	$m = 2n \equiv 0 \pmod 8$,
	Δ	*if*	$m = 2n+1 \equiv 1, 7 \pmod 8$.
(*b*) *symplectic*	: Δ^+, Δ^-	*if*	$m = 2n \equiv 4 \pmod 8$,
	Δ	*if*	$m = 2n+1 \equiv 3, 5 \pmod 8$.

First Proof. The conclusions of Theorem 4.6 are sufficient to show that $M = \Delta^+, \Delta^-, \Delta$ comes from real/symplectic representations of the finite group E_0, i.e. there is an action of $\mathbf{R}(E_0)$ on a module over $\mathbf{R}$ or $\mathbf{H}$ which affords the given representation. But $\mathbf{R}(E_0)/\langle \nu + 1 \rangle \cong Cl(V)_0$ acts on M and $Spin_{\mathbf{R}}(V) \subset Cl(V)_0$, so $Spin_{\mathbf{R}}(V)$ acts on M.

Second Proof. The argument for finite groups works for compact groups and the first sentence of the first proof suffices (with 'compact' replacing 'finite').

□

The Theorem of Weyl on $R(G)$.

We end this chapter by describing how Δ, Δ^+ and Δ^- fit into general representation theory. For example, when $G = SO(5)$, we can plot the weights of the adjoint representation as follows:

$$\begin{array}{rcl} -x_1+x_2\,\bullet & \bullet\, x_2 & \bullet\, x_1+x_2 \\ -x_1\,\bullet & \bullet\, 0 & \bullet\, x_1 \\ -x_1-x_2\,\bullet & \bullet\, -x_2 & \bullet\, x_1-x_2 \end{array}$$

The roots are the non-zero weights here and the weights of any representation are linear combinations of these. Thus if λ^i is the i-th exterior power of the standard representation of $SO(5)$ on $\mathbf{R}^5$ then the weights of various representations are as follows:

$$\lambda^1: \quad \begin{array}{ccc} & 1 & \\ 1 & 1 & 1 \\ & 1 & \end{array} \qquad\qquad \lambda^2 = adj: \quad \begin{array}{ccc} 1 & 1 & 1 \\ 1 & 2 & 1 \\ 1 & 1 & 1 \end{array}$$

where if w_i is a root (or zero) and j an integer we have replaced jw_i by j in the diagram. As a further example, we list the diagram for the second symmetric power σ^2 of the standard representation of $\mathbf{R}^5$:

$$\sigma^2 = \sigma^2(\mathbf{R}^5): \quad \begin{array}{ccccc} & & 1 & & \\ & 1 & 1 & 1 & \\ 1 & 1 & 3 & 1 & 1 \\ & 1 & 1 & 1 & \\ & & 1 & & \end{array}$$

These patterns must be invariant under the Weyl group and, for irreducible representations, all non-zero entries lie within the convex hull of the orbit of one with unit multiplicity. The irreducible representation is completely determined by giving that orbit or even one weight in it. To make this quite explicit we must choose a *fundamental dual Weyl chamber* (FDWC) where FDWC $= \{ h \in L(T)^* \mid (\theta, h) > 0 \text{ for each simple root } \theta \}$, [1, p. 125].

Let $R(G)$ be the representation ring of G (the free abelian group generated by the isomorphism classes of irreducible, complex representations of G). Then we have

Theorem 4.9. (H. Weyl) *If G is compact, connected and simply-connected then $R(G)$ is a polynomial ring with generators in 1-1 correspondence with the nodes of the Dynkin diagram as follows. For each simple root θ_i, define the corresponding edge of the FDWC by $(\phi, \theta_i) > 0$ and $(\phi, \theta_j) = 0$ for all other simple roots θ_j. Go along the edge to the first weight and take the corresponding representation s_i. This is the generator* [1, p. 164]. □

Example. $G = Spin(2n)$. Let $\lambda^i = \lambda^i(\mathbf{R}^{2n})\otimes\mathbf{C}$. Then the Dynkin diagram is

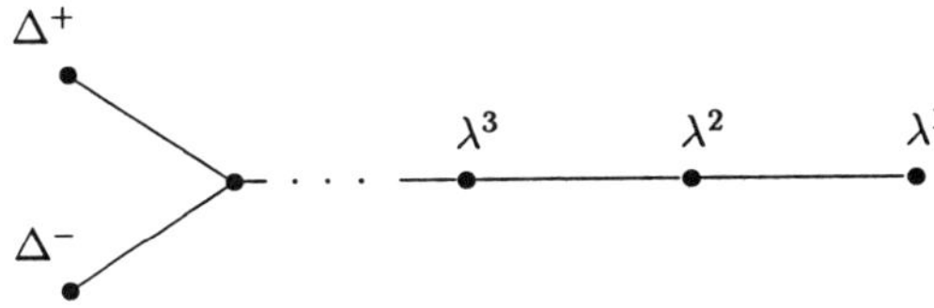

labeled as shown, so

$$\mathbf{R}(Spin(2n)) \cong \mathbf{Z}[\,\Delta^+, \Delta^-, \lambda^1, \lambda^2, \ldots, \lambda^{n-2}\,].$$

Here for example, the weight for Δ^+ is $\frac{1}{2}(x_1 + \cdots + x_n)$, corresponding to the root $x_1 + x_2$. The formulae in Theorem 4.6 are those for writing λ^{n-1}, λ^{n+}, λ^{n-} in terms of these generators.

Chapter 5

Applications of Spin representations

Construction of G_2.

For large values of n, the diagrams of type A_n, B_n, C_n and D_n are distinct. For small values of n we have the possibility of *exceptional* isomorphisms between the classical groups as follows.

(a) $Spin(6) \cong SU(4)$. Both have dimension 15, rank 3 and the Dynkin diagram  .

(b) $Spin(5) \cong Sp(2)$. Both have dimension 10, rank 2 and the Dynkin diagram  .

(c) $Spin(3) \cong SU(2) \cong Sp(1) = S^3 \subset \mathbf{H}$.

(d) $Spin(4) \cong S^3 \times S^3$.

We just prove the first two isomorphisms ; the others are easier.

Proposition 5.1. *We have the following isomorphisms of Lie groups :*

$$Spin(6) \cong SU(4), \quad Spin(5) \cong Sp(2).$$

Proof. We do both cases in parallel. $Spin(5)$ has the representation Δ of degree 4 over $\mathbf{C}$ and degree 2 over $\mathbf{H}$. We can impose a Hermitian form, invariant under the compact group $Spin(5)$, giving us a homomorphism $\Delta : Spin(5) \to Sp(2)$. Similarly $Spin(6)$ has the representation Δ^+ of degree 4 over $\mathbf{C}$ and we have $\Delta^+ : Spin(6) \to U(4)$. We first wish to show that $Im\,\Delta^+ \subset SU(4)$. Let $t \in T \subset Spin(6)$. Then t acts with eigenvalues defined by weights $\{\, \frac{1}{2}(x_1 + x_2 + x_3),\ \frac{1}{2}(x_1 - x_2 - x_3),\ \frac{1}{2}(-x_1 + x_2 - x_3),\ \frac{1}{2}(-x_1 - x_2 + x_3) \,\}$ which add to zero so the eigenvalues multiply to 1 and t must act with $\det 1$. Hence any gtg^{-1} acts with $\det 1$ and Δ^+ maps to $SU(4)$. Now Δ is faithful: if $g \in Spin(2n+1)$ and $g \mapsto 1$ in $Hom_{\mathbf{C}}(\Delta, \Delta) \cong Cl(V)_0$, then g is 1. Also Δ^+ is faithful if n is odd, for if g acts as 1 on Δ^+, then $g \in Spin(2n)$ acts as 1 on the dual, Δ^-, so $g \mapsto 1$ in $Hom_{\mathbf{C}}(\Delta^+, \Delta^+) + Hom_{\mathbf{C}}(\Delta^-, \Delta^-) = Cl(V)_0$. Thus $g = 1$. Hence the two maps Δ, Δ^+ are mono and induce monomorphisms $L(\Delta)$, $L(\Delta^+)$ and so are iso for dimensional reasons. Thus Δ and Δ^+ map small neighbourhoods in $Spin(5)$, $Spin(6)$ onto small neighbourhoods of $Sp(2)$, $SU(4)$ respectively. But $Sp(2)$ and $SU(4)$ are connected, hence $\Delta : Spin(5) \to Sp(2)$ and $\Delta^+ : Spin(6) \to SU(4)$ must be epi. □

Corollary 5.2. *The group $Spin(5)$ acts transitively on the unit sphere $S^7 \subset \Delta$.*

Proof. $Sp(2)$ acts transitively on $S^7 \subset \mathbf{H}^2$. □

Corollary 5.3. *The group $Spin(6)$ acts transitively on pairs (x, z), $x \in S^5 \subset \mathbf{R}^6$, $z \in S^7 \subset \Delta^+$. Moreover, the subgroup fixing $z \in S^7 \subset \Delta^+$ may be taken, by a suitable choice of z, to be $SU(3) \subset Spin(6)$, where this inclusion arises by lifting the composite $SU(3) \hookrightarrow U(3) \hookrightarrow SO(6)$ to $Spin(6)$.*

Proof. $Spin(6)$ covers $SO(6)$ which acts transitively on S^5. Taking a suitable x, e.g. $x = (0, \ldots, 0, 1)$, the subgroup fixing x is $Spin(5)$. The restriction of Δ^+ to $Spin(5)$ is Δ and $Spin(5)$ acts transitively on $z \in S^7 \subset \Delta$ by the last corollary. The weights of Δ^+ are $\{\frac{1}{2}(x_1 + x_2 + x_3),\ \frac{1}{2}(x_1 - x_2 - x_3),\ \ldots\}$ and their restrictions to $T \subset SU(3)$ are $\{0, x_1, x_2, x_3\}$, hence the restriction of Δ^+ to $SU(3)$ is $1 + \lambda^1$. So this $SU(3)$ fixes points of $S^7 \subset \Delta^+$: in fact, a whole circle of them. For any such fixed point, the subgroup fixing it is no bigger, since in $SU(4) \cong Spin(6)$ the subgroup fixing a unit vector in $\mathbf{C}^4$ is an $SU(3)$. □

Corollary 5.4. *The group $Spin(7)$ acts transitively on triples (x, y, z) where x, y are orthogonal in $S^6 \subset \mathbf{R}^7$ and $z \in S^7 \subset \Delta$.*

Proof. $Spin(7)$ covers $SO(7)$ which is transitive on points $y \in S^6$. Choose $y = (0, \ldots, 0, 1)$. Then the subgroup fixing y is $Spin(6)$. Now Δ restricts to $\Delta^+ + \Delta^-$ over $\mathbf{C}$, but starting with Δ as a vector space of dimension 8 over $\mathbf{R}$, the restriction to $Spin(6)$ is the representation of dimension 8 over $\mathbf{R}$ underlying Δ^+ (or Δ^-). Finally, note that by the previous corollary, $Spin(6)$ is transitive on pairs (x, z), $x \in S^5$, $z \in S^7 \subset \Delta^+$. □

Theorem 5.5. *Consider the subgroup G of $Spin(7)$ which fixes a point $z \in S^7 \subset \Delta$. Then G is a compact, connected, simply-connected, Lie group of rank 2, of dimension 14, with the Dynkin diagram*  *and commonly called G_2. Moreover, G_2 is transitive on pairs (x, y) of orthogonal vectors in $S^6 \subset \mathbf{R}^7$.*

Proof. The last sentence follows from Corollary 5.4. Clearly G is a closed subgroup of $Spin(7)$ and since $Spin(7)$ is transitive on S^7, we have $\dim G =$

$\dim Spin(7) - \dim S^7 = 21 - 7 = 14$. Let $H \subset G$ be the subgroup fixing $y = (0, \ldots, 0, 1)$. Then H is the same as the subgroup of $Spin(6)$ which fixes z and by a suitable choice of z, we can take $H = SU(3) \subset Spin(6)$. Since H is connected and $G/H = S^6$ is connected, we find that G is connected, (use the homotopy exact sequence of the fibration $H \to G \to G/H$, [S, p. 377]). Similarly G is simply-connected. Next we determine the roots of G. We know how $H = SU(3)$ acts on $L(H) \subset L(G)$, namely by the adjoint action, so next we wish to know how H acts on $L(G)/L(H)$, the tangent space to S^6 at y. By construction, $S^6 = Spin(7)/Spin(6)$, so we need to look at the action of $Spin(6)$ on $L(Spin(7))/L(Spin(6))$. We know the weights of this action, namely $\{\pm x_1, \pm x_2, \pm x_3\}$. This is the geometrically obvious action where the tangent space to S^6 at $(0, \ldots, 0, 1)$ is $\mathbf{R}^6$, the space of the first 6 coordinates and $Spin(6)$ acts on it as usual. Thus $T \subset SU(3)$ acts on $L(G)$ with weights $\{0, 0, \pm(x_1 - x_2), \pm(x_2 - x_3), \pm(x_3 - x_1), \pm x_1, \pm x_2, \pm x_3\}$. We conclude that T is maximal, these are the roots of G, and the Dynkin diagram is  . □

G_2 starts life with two obvious representations: a 7-dimensional representation $G_2 \subset Spin(7)$, acting on $\mathbf{R}^7$ with weights $\{0, \pm x_1, \pm x_2, \pm x_3\}$ and a 14-dimensional representation, *Ad* with weights as in Theorem 5.5. These are the two generators in the H. Weyl theorem (Theorem 4.9) applied to $R(G_2)$.

Spin(8) **and triality.**

We turn next to $Spin(8)$ and some background. If G is a Lie group we can form

(i) $Aut(G)$, which is another Lie group but not necessarily connected even if G is.

(ii) $Inn(G)$, the inner automorphisms given by $\alpha_g(h) = ghg^{-1}$. This is a normal subgroup of $Aut(G)$.

(iii) $Out(G) = Aut(G)/Inn(G)$.

If $\alpha \in Aut(G)$ and $\rho : G \to Aut(V)$ is a representation, then we can form $G \xrightarrow{\alpha} G \xrightarrow{\rho} Aut(V)$ and if $\alpha \in Inn(G)$ then $\rho\alpha$ is equivalent to the representation ρ. Thus $Out(G)$ permutes the equivalence classes of representations of G. In particular, $Out(Spin(8)) \cong \Sigma_3$ which permutes the three basic representations of $Spin(8)$, see Theorem 5.6 below.

Example. The $(n+1)$-dimensional representation λ^n of $SU(n+1)$ may be interpreted as $SU(n+1) \xrightarrow{\lambda^n} SU(n+1)$ and this takes λ^1 to λ^n.

Example. Define $\alpha : SO(2n) \to SO(2n)$ by $\alpha(B) = ABA^{-1}$ where $A = \begin{pmatrix} I_{2n-1} & O \\ O & -1 \end{pmatrix} \notin SO(2n)$. This α is usually called "the outer automorphism of

$SO(2n)$ which becomes inner in $SO(2n+1)$" : in $SO(2n+1)$, we can conjugate by $diag(1,\ldots,1,-1,-1)$. The resulting automorphism α lifts to $Spin(2n)$ as conjugation by $e_{2n}e_{2n+1}$ in $Spin(2n+1)$ or by e_{2n} in $Pin(2n+1)$. If T is the standard maximal torus in $Spin(2n)$, then $\alpha T = T$ and $\alpha(x_i) = x_i$ for $1 \leq i \leq n-1$, $\alpha(x_n) = -x_n$, so that α interchanges Δ^+ and Δ^-. Therefore it cannot be an inner automorphism. More generally, if $\alpha : G \to G$ is an automorphism of a compact, connected, Lie group and T a maximal torus, then αT is a torus in G, so by suitable conjugation, $\alpha T = T$ and after conjugation with a suitable element of $N(T)$, we can assume that α preserves the fundamental Weyl chamber. Thus α permutes the walls of FWC and induces an automorphism of the Dynkin diagram, which is to say that it permutes the representations attached to the nodes of the diagram.

Example. $A_n = SU(n+1)$.

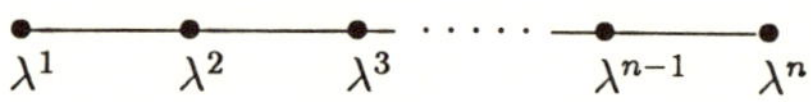

Here $\lambda^i \leftrightarrow \lambda^{n+1-i}$

D_n, $n > 4$.

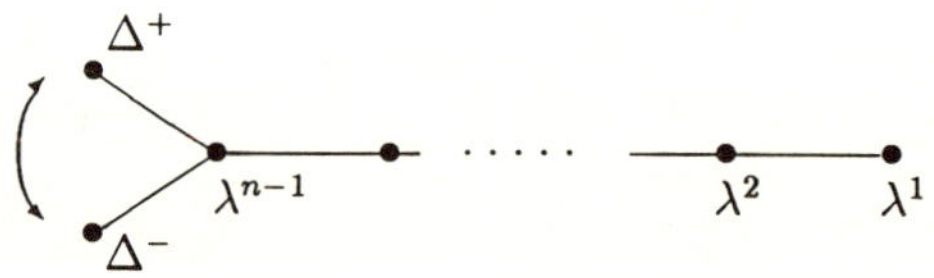

Here the roots marked $\Delta^\pm$ are interchanged, keeping the others fixed.

$D_4 = Spin(8)$.

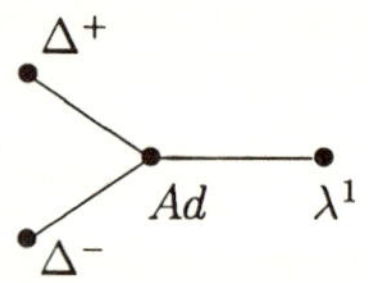

Here $\dim_{\mathbf{R}} \Delta^+ = \dim_{\mathbf{R}} \Delta^- = 8$ and G has exactly three classes of representations of degree 8, namely Δ^+, Δ^-, λ^1.

Theorem 5.6. *$Out(Spin(8)) = \Sigma_3$ acting on Δ^+, Δ^- and λ^1.*

Proof. An automorphism $\alpha : Spin(8) \to Spin(8)$ gives an 8-dimensional representations of $Spin(8)$. Conversely, given any 8-dimensional representation of $Spin(8)$ on V, we can impose a positive definite inner product, pick an orthogonal basis, get a homomorphism $Spin(8) \to O(8)$ and remark that since $Spin(8)$ is connected and simply-connected it maps into $SO(8)$ and lifts to $Spin(8)$. But the classification of automorphisms is by conjugation in $Spin(8)$ or $SO(8)$ whereas the classification of representations is by conjugation in $Pin(8)$ or $O(8)$. For example, taking $\Delta^+ : Spin(8) \to Spin(8)$, we certainly have $\Delta^+ \leftrightarrow \lambda^1$ but either

$(\lambda^1 \leftrightarrow \Delta^+,\ \Delta^- \leftrightarrow \Delta^-)$ or $(\Delta^- \leftrightarrow \Delta^+,\ \lambda^1 \leftrightarrow \Delta^-)$ and both possibilities can arise. If one arises, we get the other by composing with some $\alpha : Spin(8) \to Spin(8)$. Similarly for $\Delta^- : Spin(8) \to Spin(8)$.

Thus the map $Out(Spin(8)) \to \Sigma_3$ is onto. If some $\delta : Spin(8) \to Spin(8)$ maps to the identity in Σ_3, then $\delta^*(\lambda^1) = \lambda^1$, so δ is conjugate to 1 by an element in $Pin(8)$ of $\det \pm 1$ and since $\delta^*\Delta^+ = \Delta^+$, it must be an element of $\det 1$ so δ is inner and $Out(Spin(8)) \to \Sigma_3$ is mono, hence an isomorphism. See [6] and [5]. □

When talking of $A_n = SU(n+1)$, we call the symmetry of the diagram *duality*, $\lambda^1 \otimes \lambda^n \cong \lambda^{n+1} \cong \mathbf{C}^{n+1}$, and when talking of $Spin(8)$, we call the symmetry *triality*.

Definition. *Let V_1, V_2, V_3 be real vector spaces. Then a* triality *is a linear map $f : V_1 \otimes V_2 \otimes V_3 \to \mathbf{R}$ such that for any non-zero $v_1 \in V_1$, $v_2 \in V_2$ there exists $v_3 \in V_3$ such that $f(v_1 \otimes v_2 \otimes v_3) \neq 0$ (and similarly for $v_1, v_3 \neq 0$, $v_2, v_3 \neq 0$). If each V_i has an inner product, we say that f is a* normed triality *if $|f(v_1 \otimes v_2 \otimes v_3)| \leq ||v_1||\,||v_2||\,||v_3||$ and for all $v_1, v_2 \neq 0$ there is a $v_3 \neq 0$ for which the bound is attained (and similarly for the other two cases).*

Examples. (i) $V_1 = V_2 = V_3 = \mathbf{R}$, $\mathbf{C}$ or $\mathbf{H}$ respectively and take $f(x \otimes y \otimes z) = Re(xyz)$.

(ii) $V_1 = \Delta^+$, $V_2 = \Delta^-$, $V_3 = \lambda^1$ and f as in Proposition 5.7 below.

These examples are related as follows. Suppose given a triality f. Then for any $v_1 \neq 0$ we get a duality of V_2, V_3, so we must have $\dim V_2 = \dim V_3$ and similarly $\dim V_1 = \dim V_2 = \dim V_3$. We can transpose V_3 to get $f' : V_1 \otimes V_2 \to V_3^*$. Choose $e_1 \neq 0$ in V_1 and use it to identify V_2 with V_3^* by $v_2 \mapsto f'(e_1 \otimes v_2)$. Similarly, choose $e_2 \neq 0$ in V_2 and identify V_1 with V_3^*. We now have $f'' : V_3^* \otimes V_3^* \to V_3^*$ for which $f''(e_1 \otimes e_2)$ acts as a two-sided unit and f'' is non-singular in that if $x, y \neq 0 \in V_3^*$, then $f''(x \otimes y) \neq 0$. So we have a division algebra over $\mathbf{R}$ and consequently $\dim V_i = 1, 2, 4$ or 8. If we start with a normed triality, we get a normed algebra.

In example (i), for $\mathbf{R}$, the symmetry group of this triality is $\dfrac{S^0 \times S^0 \times S^0}{S^0} = S^0 \times S^0$; for $\mathbf{C}$ the symmetry group is $\dfrac{S^1 \times S^1 \times S^1}{S^1} = T^2$; for $\mathbf{H}$ the group is $\dfrac{S^3 \times S^3 \times S^3}{\{\pm 1\}}$.

Proposition 5.7. *There is a non-zero trilinear map $f : \Delta^+ \otimes \Delta^- \otimes \lambda^1 \to \mathbf{C}$ invariant under $Spin(8)$ and unique up to a non-zero scalar multiple. If we work with real representations, we may insist that f be real valued and for $x \in S^7 \subset \Delta^+$, $y \in S^7 \subset \Delta^-$, $z \in S^7 \subset \lambda^1 = \mathbf{R}^8$ we have*

$$-1 \leq f(x \otimes y \otimes z) \leq 1$$

where both bounds are attained. The map f is then unique up to a factor ± 1.

Proof. By Theorem 4.6 , we have $\Delta^+ \otimes \Delta^- \cong \lambda^1 + \lambda^3$. Here λ^1 and λ^3 are irreducible, λ^1 is dual to λ^1 but not dual to λ^3. Therefore

$$\lambda^1 \otimes \Delta^+ \otimes \Delta^- = \lambda^1 \otimes \lambda^1 + \lambda^1 \otimes \lambda^3 \cong Hom_{\mathbf{C}}(\lambda^1, \lambda^1) + Hom_{\mathbf{C}}(\lambda^1, \lambda^3).$$

Only the first summand maps $Spin(8)$-invariantly onto $\mathbf{C}$, proving the first part of the proposition. But $S^7 \times S^7 \times S^7$ is compact, so putting

$$c = \max_{||x||=||y||=||z||=1} |f(x \otimes y \otimes z)|$$

we have $-c \leq f(x \otimes y \otimes z) \leq c$. The bounds are attained because we can change the signs of x, y and z. Also $c > 0$ since $f \neq 0$. By multiplying f by a non-zero constant, we can assume $c = 1$ and this is attained. $\square$

Chapter 6

The exceptional groups: construction of E_8

We draw the following lesson from our construction of G_2: To describe an unknown group G, it is useful to find a known subgroup of maximal rank $H \subset G$ and to give an account of G/H.

Theorem 6.1. *There exist Lie groups G with subgroups H as specified in the following table.*

local type of H	rank	dim	$L(G)/L(H)$ as $\mathbf{C}$ rep.	dim	G	rank	dim
$Spin(9)$	4	36	Δ	16	F_4	4	52
$Spin(10)\times U(1)/\mathbf{Z}_4$	6	46	$\Delta^+\otimes\xi^3+\Delta^-\otimes\xi^{-3}$	32	E_6	6	78
$Spin(12)\times Sp(1)/\mathbf{Z}_2$	7	69	$\Delta^+\otimes\lambda^1$	64	E_7	7	133
$Spin(16)/\mathbf{Z}_2$	8	120	Δ^+	128	E_8	8	248

Notes. 1. In $Spin(16)/\mathbf{Z}_2$ the $\mathbf{Z}_2$ is generated by $\prod_1^{16} e_i$ which acts on Δ^+ as $i^8 = 1$ and generates $\ker \Delta^+$.

2. In $Spin(12) \times Sp(1)/\mathbf{Z}_2$ the $\mathbf{Z}_2$ is generated by $(\prod_1^{12} e_i, -1)$ which acts on $\Delta^+ \otimes \lambda^1$ as $i^6 \cdot (-1) = 1$.

3. In $Spin(10) \times U(1)/\mathbf{Z}_4$ the $\mathbf{Z}_4$ is generated by $(\prod_1^{10} e_j, i)$ which acts on $\Delta^+ \otimes \xi^3 + \Delta^- \otimes \xi^{-3}$ as $i^5 \otimes i^3$.

The first column to be filled is that for $\dim(L(G)/L(H))$ and then we proceed to find groups which have representations Δ or Δ^+ of these dimensions. We begin with the construction of the Lie algebra of E_8, look for subgroups in the next chapter and finish the proof of Theorem 6.1 in Chapter 8.

Construction of a Lie algebra of type E_8.

For E_8, there is no representation of smaller degree than Ad, so we use that. Take $L + \Delta^+$ where $L = L(Spin(16))$ and consider this simultaneously over **R** and **C**. For a while we can work with $Spin(2n)$. Now, by Proposition 3.1, $L \subset Cl(2n)_0$ has a basis $\{ e_r e_s \mid r < s \}$, and Δ^+ is a representation of $Spin(2n)$ or L, over **R**. That is, for all $a \in L$, $u \in \Delta^+$ we have $[a,u] \in \Delta^+$ satisfying the Jacobi identity $[[a,b],u] = [a,[b,u]] - [b,[a,u]]$ and in fact $[a,u]$ is just the usual multiplication of $u \in \Delta^+ = \mathbf{C}(E_0) \otimes_{\mathbf{C}(F)} W$ (Proposition 4.2) by $a \in Cl(2n)_0$. Assume now that $2n \equiv 0 \pmod 8$ and consider Δ^+ as a real representation of $Spin(2n)$. Choose $\mu : \Delta^+ \otimes \Delta^+ \to \mathbf{R}$, a symmetric, bilinear, non-zero map invariant under $Spin(2n)$, i.e. $(gu, gv) = (u,v)$ for $g \in Spin(2n)$, $u, v \in \Delta^+$, see [1, p. 36]. The linearised form of this is: $([a,u],v) + (u,[a,v]) = 0$. (This is the equation to be used whenever we say that a bilinear form is invariant under a Lie algebra.) The elements of L can be considered as skew-symmetric matrices (see the proof of Proposition 3.1) and on matrices we have an inner product $(A,B) = Tr(AB)$ which is a symmetric bilinear form, real on real matrices. The invariance property for $X \in GL(2n)$ is $Tr(XAX^{-1}XBX^{-1}) = Tr(AB)$. If we put $X = I + tY$ and pass to the limit, we obtain the linearised version, $Tr([Y,A]B + A[Y,B]) = 0$, so we get $([Y,A],B) + (A,[Y,B]) = 0$. Under this identification, $e_r e_s$ corresponds to the matrix with all entries zero except in positions (r,s) and (s,r) where we have respectively -2 and 2. This gives $(e_r e_s, e_r e_s) = -8$ and to remove this undesirable factor, we put $(A,B) = -\frac{1}{8}Tr(AB)$ so that now $(e_r e_s, e_t e_u) = \delta_{rt}\delta_{su}$. In the next lemma, we transpose the action $L \otimes \Delta^+ \to \Delta^+$ above to get a map $\Delta^+ \otimes \Delta^+ \to L$.

Lemma 6.2. *For all $u, v \in \Delta^+$, there is a unique $[u,v] \in L$ such that $(a,[u,v])_L = ([a,u],v)_{\Delta^+}$ for all $a \in L$ and $[u,v]$ is bilinear in u, v. Furthermore, if $v, w \in \mathbf{C} \otimes \Delta^+$ are such that*

$$e_{2q-1}e_{2q}v = iv \quad \text{for all } q \quad (\text{corresponding to a weight } \tfrac{1}{2}\sum_1^n x_i),$$

$$e_{2q-1}e_{2q}w = -iw \quad \text{for all } q \quad (\text{corresponding to a weight } -\tfrac{1}{2}\sum_1^n x_i),$$

and $(v,w) = 1$, then

(i) $[v,w] = i(e_1e_2 + e_3e_4 + \cdots + e_{2n-1}e_{2n})$;

(ii) $[e_{2q}e_{2r}v, w] = (e_{2q-1} + ie_{2q})(e_{2r-1} + ie_{2r}), \quad q < r$;

(iii) $[e_{2q_1}e_{2q_2}\dots e_{2q_{2m}}v, w] = 0$ *if* $m > 1$ *and* $q_1 < q_2 < \cdots < q_{2m}$.

Proof. Clearly $([a,u],v)$ is a linear function of a and since the inner product on L is non-singular, we must have $([a,u],v) = (a,b)$ for some $b = [u,v] \in L$. Since $([a,u],v)$ is bilinear in u, v, so is b. Everything so far works over **R** or **C**. We proceed to derive the explicit formulae.

(i) First we have $(e_{2q-1}e_{2q}v, w) = (iv, w) = i$ and $(e_r e_s v, w) = 0$ if $e_r e_s$ is not one of the basis elements $e_{2q-1}e_{2q}$. Thus $[v,w]$ is paired to i if $a = e_{2q-1}e_{2q}$ and to 0 for all other basis elements.

(ii) For simplicity of notation consider $[e_2e_4v, w]$. Then $(e_re_se_2e_4v, w) = 0$ except when $(r,s) = (1,3),(1,4),(2,3)$, or $(2,4)$, when we get respectively $1, i, i, -1$. This yields $[e_2e_4v, w] = e_1e_3 + ie_1e_4 + ie_2e_3 - e_2e_4$. Notice that e_2e_4v has weight $\frac{1}{2}(-x_1 - x_2 + x_3 + \cdots + x_n)$ while w has weight $\frac{1}{2}(-x_1 - \cdots - x_n)$ so that $[e_2e_4v, w]$ must have weight $-x_1 - x_2$. In fact, $e_1 + ie_2$ has weight $-x_1$ and $e_3 + ie_4$ has weight $-x_2$.

(iii) It suffices to note that for all e_re_s, $r < s$, $(e_re_se_{2q_1}e_{2q_2}\cdots e_{2q_{2m}}v,\ w) = 0$.

□

Remark. The map $[\ ,\ ] : \Delta^+ \otimes \Delta^+ \to L$ is invariant under $Spin(2n)$ because everything in the construction is invariant under $Spin(2n)$. The linearised form of invariance (that is invariance under L) is

$$[a, [u, v]\,] = [\,[a, u], v] + [u, [a, v]\,]$$

and this equation is established as follows.

It is sufficient to show that for all $b \in L$

$$(-b, [a, [u, v]\,]\,) + (b, [\,[a, u], v]\,) + (b, [u, [a, v]\,]\,)$$

is zero. This expression, using the invariance of $(\ ,\ )_L$ under L and the definition of $[u, v]$, is

$$(\,[a, b], [u, v]\,) + (\,[b, [a, u]\,], v) + (\,[b, u], [a, v]\,).$$

Using the invariance of $(\ ,\ )_{\Delta^+}$, this is

$$(\,[\,[a, b], u], v) + (\,[b, [a, u]\,], v) - (\,[a, [b, u]\,], v) = 0,$$

using properties of the action of L on Δ^+. □

We now proceed to give $L + \Delta^+$ the inner product with L and Δ^+ orthogonal and

$$(a + u, b + v) = (a, b)_L + (u, v)_{\Delta^+}, \quad \text{for all } a, b \in L,\ u, v \in \Delta^+.$$

The Lie bracket $[a, b]$ is as in L, $[a, u]$ as the action of L on Δ^+, $[u, a] = -[a, u]$, and $[u, v]$ as in Lemma 6.2.

Theorem 6.3. *If $2n = 16$, $L + \Delta^+$ becomes a Lie algebra with an invariant inner product.*

Proof. The inner product is invariant under L by definition, and under Δ^+ by the definition of $[u, v]$. We need to prove anti-commutativity and the Jacobi identity. Clearly $[a, b] = -[b, a]$ since L is a Lie algebra, and we define $[u, a]$ to be $-[a,u]$. To see that $[u,v] = -[v,u]$ observe that for all $a \in L$, $u,v \in \Delta^+$, we have

$$\begin{aligned}(a,[u,v]+[v,u]) &= ([a,u],v) + ([a,v],u) \\ &= ([a,u],v) + (u,[a,v]), \quad \text{by symmetry of } (\ ,\ ), \\ &= 0, \quad \text{by the invariance of } (\ ,\ ) \text{ under } a \in L.\end{aligned}$$

Since this is true for all a, we have $[u,v]+[v,u] = 0$.

For the Jacobi identity, we need to discuss several cases.

(i) Three variables in L, none in Δ^+: The identity will hold here since L is a Lie algebra.

(ii) Two variables in L, one in Δ^+: We have $[[a,b],u] = [a,[b,u]] - [b,[a,u]] = 0$. Thus $[a,[b,u]] + [u,[a,b]] + [b,[u,a]] = 0$.

(iii) One variable in L, two in Δ^+: Here, by the invariance of the bracket under L, $[a,[u,v]] = [[a,u],v] + [u,[a,v]]$, which leads to the Jacobi identity.

(iv) All three variables in Δ^+: This is where we have to use the fact that $n = 8$. We set down a general procedure for checking identities of this type by using symmetry.

Standard operating procedure.

Step a. Express the identity to be verified as the vanishing of a G-map $f : U \to V$ where U, V are representations of the compact, connected, Lie group G.

Example. $G = Spin(16)$, $f : \Delta^+ \otimes \Delta^+ \otimes \Delta^+ \to \Delta^+$, $x \otimes y \otimes z \mapsto [x,[y,z]] + [y,[z,x]] + [z,[x,y]]$.

Step b. It is sufficient to check that $f = 0$ on a basis for U consisting of eigenvectors of the T-action where $T \subset G$ is a maximal torus.

Step c. If $f(u) = 0$ for one eigenvector $u \in U$, then the result follows for each nu, $n \in N(T)$. (Proof: $f(nu) = nf(u) = 0$ and nu is an eigenvector since $t(nu) = n(n^{-1}tn)u = nt'u = n\lambda(t)u = \lambda(t)nu$.) We need only check on eigenvectors corresponding to one weight in each orbit of the Weyl group.

Thus it is sufficient for us to check the Jacobi identity when x, y, z are eigenvectors of T. Suppose

$$\begin{array}{lll} x \text{ has weight} & \frac{1}{2}\sum \alpha_r x_r, & \alpha_r = \pm 1, \\ y & \frac{1}{2}\sum \beta_r x_r, & \beta_r = \pm 1, \\ z & \frac{1}{2}\sum \gamma_r x_r, & \gamma_r = \pm 1. \end{array}$$

Thus $f(x \otimes y \otimes z)$ is an eigenvector with weight $\frac{1}{2}\sum_{r=1}^{8}(\alpha_r + \beta_r + \gamma_r)x_r$ and we can assume $f(x \otimes y \otimes z) = 0$ unless $\frac{1}{2}\sum(\alpha_r + \beta_r + \gamma_r)x_r$ is one of the weights,

$\frac{1}{2}\sum \varepsilon_r x_r$, $\varepsilon_r = \pm 1$, which occur in Δ^+, that is, unless for all r, two of α_r, β_r, γ_r are ε_r and one is $-\varepsilon_r$. We now have four cases to consider.

Case 1. One of x, y, z has all 8 occurrences of $-\varepsilon_r$.

Case 2. There are 6 occurrences of $-\varepsilon_r$ in one of x, y, z, 2 in another, 0 in the third.

Case 3. There are 4 occurrences of $-\varepsilon_r$ in one of x, y, z, 4 in another, 0 in the third.

Case 4. There are 4 occurrences of $-\varepsilon_r$ in one of x, y, z, 2 in another, 2 in the third.

These are all the possibilities since the Weyl group permutes $x_1, \ldots, x_8$ in any way and changes an even number of signs. In any case, it does not matter which of the three vectors is x, y or z because $f(x \otimes y \otimes z)$ is anti-symmetric in x, y, z by the anti-symmetry of $[\ ,\ ]$.

Case 1. We may assume that all eight coefficients $-\varepsilon_r$ occur in z, so $x = y$. The result follows as f is anti-symmetric in x, y. We now deal with cases 2, 3, 4.

Case 2. By permuting the x_i, we can assume the 2 occurrences come from x_1, x_2 and by changing an even number of signs, we are reduced to

$$\begin{array}{rclll}
\varepsilon_i & = & - \ - \ + \ + \ + \ + \ + \ + & \\
\alpha_i & = & - \ - \ - \ - \ - \ - \ - \ - & \text{corresponding to } w \text{ in Lemma 6.2} \\
\beta_i & = & + \ + \ + \ + \ + \ + \ + \ + & \text{corresponding to } v \\
\gamma_i & = & - \ - \ + \ + \ + \ + \ + \ + & \text{corresponding to } e_2 e_4 v.
\end{array}$$

Case 3. By similar permutations, we can assume the four occurrences to be in x_1, x_2, x_3, x_4 and with sign changes, we have

$$\begin{array}{rclll}
\varepsilon_i & = & - \ - \ - \ - \ + \ + \ + \ + & \\
\alpha_i & = & - \ - \ - \ - \ - \ - \ - \ - & \text{corresponding to } w \\
\beta_i & = & + \ + \ + \ + \ + \ + \ + \ + & \text{corresponding to } v \\
\gamma_i & = & - \ - \ - \ - \ + \ + \ + \ + & \text{corresponding to } e_2 e_4 e_6 e_8 v.
\end{array}$$

Case 4. We assume the 2,2,4 occurrences are in $x_1 x_2$, $x_3 x_4$, $x_5 x_6 x_7 x_8$ and then we have, with sign changes,

$$\begin{array}{rclll}
\varepsilon_i & = & - \ - \ - \ - \ + \ + \ + \ + & \\
\alpha_i & = & - \ - \ - \ - \ - \ - \ - \ - & \text{corresponding to } w \\
\beta_i & = & - \ - \ + \ + \ + \ + \ + \ + & \text{corresponding to } e_2 e_4 v \\
\gamma_i & = & + \ + \ - \ - \ + \ + \ + \ + & \text{corresponding to } e_6 e_8 v.
\end{array}$$

We check the identity in each of the remaining cases.

Case 2. $[v,w] = i(e_1e_2 + \cdots + e_{15}e_{16})$, by Lemma 6.2, so

$$[\,[v,w], e_2e_4v] = i(e_1e_2 + \cdots + e_{15}e_{16})e_2e_4v = i(e_2e_4)(-2+6)iv = -4e_2e_4v.$$

$$\begin{aligned}[\,[e_2e_4v,w],v] &= (e_1+ie_2)(e_3+ie_4)v = (e_1+ie_2)(e_4e_3e_4v + ie_4v)\\ &= (e_1+ie_2)(2ie_4v) = -2ie_1e_2e_3e_4v - 2e_2e_4v = -4e_2e_4v.\end{aligned}$$

Since the weight of $[v, e_2e_4v]$ does not occur in Δ^+, $[v, e_2e_4v] = 0$. Thus

$$[\,[v,w], e_2e_4v] + [\,[w, e_2e_4v], v] + [\,[e_2e_4v, v], w] = 0.$$

Case 3. Here,

$$\begin{aligned}[\,[v,w], e_2e_4e_6e_8v] &= i(e_1e_2 + \cdots + e_{15}e_{16})e_2e_4e_6e_8v\\ &= i(e_2e_4e_6e_8)(-4+4)iv = 0, \qquad \text{and}\\ [e_2e_4e_6e_8v, w] = 0, &\qquad [e_2e_4e_6e_8v, v] = 0,\end{aligned}$$

so the identity holds.

Case 4.

$$\begin{aligned}[e_2e_4v, w] &= (e_1+ie_2)(e_3+ie_4)v, \qquad \text{so}\\ [\,[e_2e_4v,w], e_6e_8v] &= (e_1+ie_2)(e_3+ie_4)e_6e_8v\\ &= e_6e_8(-4e_2e_4v) = -4e_2e_4e_6e_8v,\\ [\,[e_6e_8v,w], e_2e_4v] &= -4e_2e_4e_6e_8v, \quad \text{and} \quad [e_2e_4v, e_6e_8v] = 0\end{aligned}$$

because the weight is not in L. Thus

$$[\,[e_2e_4v,w], e_6e_8v] + [\,[w, e_6e_8v], e_2e_4v] + [\,[e_6e_8v, e_2e_4v], w] = 0,$$

and we have shown that indeed, with $n = 8$, $L + \Delta^+$ is a Lie algebra with invariant inner product. This finishes the proof of Theorem 6.3. □

The Killing form.

For any two elements x, y of a Lie algebra A, the map $A \to A$ given by $z \mapsto [x,[y,z]\,]$ is linear, so we may set $(x,y)_K = Tr(z \mapsto [x,[y,z]\,])$, the *Killing form* of A.

Remarks 1. The Killing form is symmetric and bilinear.

Proof. It is clearly bilinear. If we define $\alpha, \beta : A \to A$ by $\alpha(a) = [x,a]$, $\beta(a) = [y,a]$, then $(x,y)_K = Tr(\alpha\beta) = Tr(\beta\alpha) = (y,x)_K$. □

2. The Killing form is invariant under A, for if we define $\gamma : A \to A$ by $\gamma(a) = [z,a]$, then, using the Jacobi identity, we have that taking Lie product with $[x,y]$ is the map $\alpha\beta - \beta\alpha$ and with $[x,z]$ is the map $\alpha\gamma - \gamma\alpha$. Thus

$$(\,[x,y],z)_K + (y,[x,z]\,)_K = Tr(\alpha\beta\gamma - \beta\alpha\gamma + \beta\alpha\gamma - \beta\gamma\alpha) = Tr(\alpha\beta\gamma - \beta\gamma\alpha) = 0.$$

3. Clearly the Killing form of an abelian Lie algebra is zero.

Lemma 6.4. *The Killing form on* $L + \Delta^+$ *is non-singular: indeed,*

$$(x, y)_K = -240(x, y).$$

Proof. Both $(\ ,\ )_K$ and $(\ ,\ )$ are invariant under $Spin(16)$ and L and Δ^+ are irreducible representations of $Spin(16)$ which are not dual to one another. We can define a map $f : A \to A$, where $A = L + \Delta^+$, by $(x, y)_K = (fx, y)$ for all $y \in A$ and then A splits as a sum of eigenspaces of f invariant under $Spin(16)$. Thus $(a+u, b+v)_K = \lambda(a, b) + \mu(u, v)$. But $(\ ,\ )_K$ and $(\ ,\ )$ are invariant under A and A is an irreducible representation of A, for the only possible subspaces closed under L are L and Δ^+ and they are not closed under Δ^+. Thus $\lambda = \mu$, and to find λ we calculate $(e_1e_2, e_1e_2)_K = Tr(z \mapsto [e_1e_2, [e_1e_2, z]]) = Tr(\delta)$, say. Now, $[e_1e_2, e_1e_2] = 0$, $[e_1e_2, e_1e_r] = 2e_2e_r$, $[e_1e_2, e_2e_r] = -2e_1e_r$ and $[e_1e_2, e_re_s] = 0$, for $2 < r < s$. So $\delta(e_1e_2) = 0$, $\delta(e_1e_r) = -4e_1e_r$, $\delta(e_2e_r) = -4e_2e_r$, $\delta(e_re_s) = 0$ and $Tr_L\delta = -112$. On Δ^+ the action is Clifford multiplication, so $[e_1e_2, [e_1e_2, u]] = e_1e_2e_1e_2u = -u$, so $Tr_{\Delta^+}(\delta) = -128$ and we have $Tr_A(\delta) = -240$. Hence $\lambda = -240$ since $(e_1e_2, e_1e_2) = 1$. □

Chapter 7

Construction of a Lie group of type E_8

Our construction of the simple, connected, compact Lie group with Lie algebra of type E_8 constructed in Chapter 6, proceeds according to the following steps.

Step 1. Take the Lie algebra $L+\Delta^+$ (over $\mathbf{R}$ or $\mathbf{C}$), where $L = L(Spin(16))$.

Step 2. Take the group of automorphisms of this Lie algebra: this is the closed subgroup of $GL(L+\Delta^+)$ preserving the Lie bracket.

Step 3. Take the identity component and call it G. (In fact the result of Step 2 is already connected.)

All our constructions are invariant under $Spin(16)$, over $\mathbf{R}$ or $\mathbf{C}$, so we get a map $Spin(16) \to Aut(L+\Delta^+)$ and since $Spin(16)$ is connected, we get a homomorphism into G. To find the kernel, note that $e_1e_2\cdots e_{16} \in Spin(16)$ acts as $i^8 = 1$ on Δ^+. It covers $-I \in SO(16)$, so it acts as -1 on $\mathbf{R}^{16}$ but it acts as 1 on L. Therefore it acts as 1 on $L+\Delta^+$. This and the identity are the only elements which act as 1 on $L+\Delta^+$, so we get an embedding $Spin(16)/\mathbf{Z}_2 \to G$. Now we proceed to check that G has the required properties.

Let A be a finite dimensional algebra over $\mathbf{R}$ or $\mathbf{C}$ (for example, a Lie algebra) and let $Aut(A)$ be the group of automorphisms of A, that is, linear bijections $\alpha : A \to A$ such that $\alpha(ab) = \alpha(a)\alpha(b)$. Then $Aut(A)$ is a closed subgroup of $GL(A)$, hence a Lie group.

Definition. *A linear map* $\partial : A \to A$ *is a* derivation *if* $\partial(ab) = (\partial a)b + a\partial(b)$.

Lemma 7.1. *$L(Aut(A))$ is the algebra of derivations of A.*

Note that the commutator $[\partial, \partial']$ is a derivation if ∂, ∂' are, so the derivations form a Lie algebra, $Der(A)$. We have a map $\partial : A \to Der(A)$ defined as follows: $x \in A$ maps to ∂_x, where $\partial_x(y) = [x, y]$. One checks that ∂_x is a derivation and ∂ a map of Lie algebras. The next lemma says that ∂ is an isomorphism if the Killing form is non-singular.

Lemma 7.2. *If A is a Lie algebra with non-singular Killing form then its algebra of derivations is A.*

Corollary 7.3. *With the above constructions, $L(G) = L + \Delta^+$.*

Proof. Clear from Lemma 6.4, Lemma 7.2 and Lemma 7.1. □

Proof of Lemma 7.1. First, $L(Aut(A)) \subset Der(A)$. To see this, take a short curve in $Aut(A)$, $\alpha_t = 1 + \gamma t + o(t^2)$, starting at the identity. Thus $\alpha_t(ab) = \alpha_t(a)\alpha_t(b)$ which gives us $\gamma(ab) = \gamma(a)b + a\gamma(b)$, so $\gamma \in L(Aut(A))$ is a derivation.

Conversely, we show that $Der(A) \subset L(Aut(A))$. If $\delta : A \to A$ is a derivation, we have by induction $\delta^n(ab) = \sum_{i+j=n} \binom{n}{i}(\delta^i a)(\delta^j b)$. Define $\alpha_t : A \to A$ by $\alpha_t = \sum_{n=0}^{\infty} \frac{t^n}{n!}\delta^n$. Then $\alpha_t(ab) = \sum_{i,j} \frac{t^{i+j}}{i!\,j!}(\delta^i a)(\delta^j b) = (\alpha_t a)(\alpha_t b)$, so $\alpha_t \in Aut(A)$, and the tangent vector to α is $\delta \in L(Aut(A))$. □

Proof of Lemma 7.2 (Zassenhaus). To see that ∂ is mono, suppose $x \in A$ and $\partial_x = 0$, i.e. $[x, y] = 0$ for all y. Then $[w, [x, y]] = 0$ for all y, i.e. $[w, [x, -]]$ is the zero function so $Tr[w, [x, -]] = 0$, i.e. $(w, x)_K = 0$ for all w, so $x = 0$ as $(\ ,\ )_K$ is non-singular.

To see that ∂ is onto, let us identify ∂_x with x and extend the Killing form from elements x to derivations by setting $(\delta, \varepsilon)_K = Tr(\delta\varepsilon : A \to A)$. We check that $([\gamma, \delta], \varepsilon)_K + (\delta, [\gamma, \varepsilon])_K = 0$ so the extension is invariant. Notice that the element $\delta(y)$ should be identified with the derivation $[\delta, \partial_y]$, since $[\delta, \partial_y](z) = \delta(\partial_y z) - \partial_y(\delta z) = \delta[y, z] - [y, \delta z] = [\delta y, z] + [y, \delta z] - [y, \delta z] = [\delta y, z]$. Thus, on substituting γ for $x = \partial_x$ and ε for z, the above invariance becomes $(-\delta x, z)_K + (\delta, [x, z])_K = 0$.

Lemma. *If $(\delta, y)_K = 0$ for all y then $\delta = 0$.*

Proof. Invariance, $(-\delta x, z)_K + (\delta, [x, z])_K = 0$, leads to $(-\delta x, z)_K = 0$ for all z in A and since the Killing form is non-singular, we have $-\delta x = 0$ for all x, that is, $\delta = 0$. □

To continue our proof that $\partial : A \to Der(A)$ is onto, take $\delta \in Der(A)$. Then $(\delta, y)_K$ is a linear function in y so has the form $(x, y)_K$ for some $x \in A$, as the Killing form is non-singular. Thus $(\delta - \partial_x, y)_K = 0$ for all y, so by the above Lemma $\delta - \partial_x = 0$ and $\delta = \partial_x$ for some x. (Cf. [10, p. 74].) □

Real Forms of E_8.

If this construction is performed over **R** and the inner product on Δ^+ is taken to be positive definite, then we have

Proposition. *The group, G, obtained is compact.*

Proof. The inner product on L is positive definite, so also positive on $L+\Delta^+$, and G preserves it. Thus G is a closed subgroup of $O(248)$ and hence compact. $\square$

With the same details, consider the subspace $L+i\Delta^+ \subset \mathbf{C}\otimes(L+\Delta^+)$. This is a real vector space with the same complexification as before. Also for all $a, b \in L$, $u, v \in \Delta^+$ we have $[a,b] \in L$, $[a, iu] \in i\Delta^+$, $[iu, iv] = -[u,v] \in L$, so we have a Lie algebra over $\mathbf{R}$ again, which is not isomorphic as a real Lie algebra to $L+\Delta^+$. To see this, look at the signatures of the Killing forms. For the compact version, the Killing form is negative definite, with signature -248. For the other version, $(\ ,\ )_K$ is negative definite on L, $\dim L = 120$ and positive definite on $i\Delta^+$, $\dim(i\Delta^+) = 128$, so the signature of $(\ ,\ )_K$ is 8. The second case is exactly the same as if we had taken the inner product on Δ^+ to be negative definite in our construction.

Chapter 8

The construction of Lie groups of types F_4, E_6, E_7

In this chapter we finally finish the discussion of Theorem 6.1, beginning with the construction of the remaining exceptional Lie groups and then the representations. We also complete our treatment of the real forms of E_8.

Step 1. The usual map $Spin(12) \times Spin(4) \to Spin(16)$ leads to

$$\begin{array}{ccccccc} & & Spin(4) & \longrightarrow & Spin(12) \times Spin(4) & \longrightarrow & Spin(16) \longrightarrow E_8 \\ & \nearrow & \downarrow & & & & \\ SU(2) \longrightarrow U(2) & \longrightarrow & SO(4). & & & & \end{array}$$

Step 2. Take the centraliser of the image of this $SU(2)$ in E_8.

Step 3. Take the identity component and call it E_7.

Note. If we work with the compact versions of the groups, we get E_7 as a closed subgroup of E_8, hence a compact group. If we work with the complex versions of the subgroups inside the complex version of E_8, we get the complex version of E_7.

Step 4. The map $Spin(10) \times Spin(6) \to Spin(16)$ gives

$$\begin{array}{ccccc} & & Spin(6) & \longrightarrow & Spin(16) \longrightarrow E_8 \\ & \nearrow & \downarrow & & \\ SU(3) \longrightarrow U(3) & \longrightarrow & SO(6). & & \end{array}$$

Step 5. Take the centraliser of the image of this $SU(3)$ in E_8.

Step 6. Take the identity component and call it E_6. We now have

$$E_6 \subset E_7 \subset E_8.$$

Step 7. The map $Spin(9) \times Spin(7) \to Spin(16)$ gives

$$G_2 \hookrightarrow Spin(7) \longrightarrow Spin(16) \longrightarrow E_8.$$

Step 8. Take the centraliser of G_2 in E_8.

Step 9. Take the identity component and call it F_4.

Remark. The embedding $G_2 \subset Spin(7)$ depends on the choice of a point in $S^7 \subset \Delta$. In Theorem 5.5, we had $SU(3) \subset G_2$ embedded using the first six coordinates. On the other hand, if the embedding $SU(3) \subset G_2$ is by the last six coordinates, we will get $F_4 \subset E_6$. Then we have

$$\left.\begin{array}{l} SU(2) \subset SU(3) \subset G_2 \\ F_4 \subset E_6 \subset E_7 \end{array}\right\} \subset E_8.$$

Identifications of the subgroups H.

We now proceed to check the subgroups H as in Theorem 6.1.

1. Take F_4 **first.**

From the sequence

$$G_2 \hookrightarrow Spin(9) \times Spin(7) \longrightarrow Spin(16) \longrightarrow E_8,$$

it is clear that $Spin(9)$ centralises the whole of $Spin(7)$ and hence it centralises $G_2 \subset Spin(7)$. One can check that $Spin(9) \to E_8$ is mono, so that F_4 has a subgroup $Spin(9)$.

2. Next take E_7.

We have the diagram

$$\begin{array}{c} Spin(12) \times Spin(4) \longrightarrow Spin(16) \longrightarrow E_8 \\ \Vert \\ (Spin(12) \times S^3) \times SU(2). \end{array}$$

The embedding of the two copies of S^3 is as $\ker \Delta^+$ and $\ker \Delta^-$; we take the embedding of $SU(2)$ as $\ker \Delta^+$. We see that $Spin(12) \times S^3$ centralises $SU(2) \cong S^3 \cong Sp(1)$ and hence that we have a map $Spin(12) \times S^3 \to E_7$. To find the kernel of this, consider $Spin(16) \to E_8$ which has kernel $\mathbf{Z}_2$ generated by $\prod_{i=1}^{16} e_i$; the kernel of $Spin(12) \times Spin(4) \to Spin(16)$ is $\mathbf{Z}_2$ generated by $(-1,-1)$. Therefore $Spin(12) \times Spin(4) \to E_8$ has kernel generated by $(e_1 \ldots e_{12},\ e_{13} \ldots e_{16})$ and $(-1,-1)$. But $e_{13} \ldots e_{16}$ lies in $\ker \Delta^-$ and not in $\ker \Delta^+$ whereas $-e_{13} \ldots e_{16}$ is in $\ker \Delta^+$ and not in $\ker \Delta^-$. Hence only the $\mathbf{Z}_2$ generated by $(e_1 \ldots e_{12},\ e_{13} \ldots e_{16})$ lies in $Spin(12) \times S^3 \subset E_7$. Hence $Spin(12) \times S^3/\mathbf{Z}_2 \to E_7$ is mono.

3. Finally, take E_6.

Consider the diagrams

$$Spin(10) \times Spin(6) \longrightarrow Spin(16) \longrightarrow E_8$$

and

$$\begin{array}{ccccccc} (z,g) & & S^1 \times SU(3) & & \longrightarrow & & Spin(6) \\ \downarrow & & \downarrow & & & & \downarrow \\ (z^2,g) & & S^1 \times SU(3) & \longrightarrow & U(3) & \longrightarrow & SO(6) \end{array}$$

where the maps are the obvious ones.

We have $Spin(10) \times S^1 \times SU(3) \to Spin(16) \to E_8$ where $Spin(10) \times S^1$ centralises the (image of) $SU(3)$ in E_8. It remains to check the kernel of $Spin(10) \times S^1 \to E_8$. The kernel of $Spin(16) \to E_8$ is $\mathbf{Z}_2$ generated by $e_1 e_2 \dots e_{16}$ and the kernel of $Spin(10) \times Spin(6) \to Spin(16)$ is $\mathbf{Z}_2$ generated by $(-1,-1)$. Thus $\ker(Spin(10) \times Spin(6) \to E_8)$ has four elements, generated by $(e_1 \dots e_{10}, e_{11} \dots e_{16})$ and $(-1,-1)$, so this kernel is $\mathbf{Z}_4$, with all four elements in S^1. (To see this, note that S^1 is the image of $t \mapsto \big(\cos t + (\sin t)e_{11}e_{12}\big)\big(\cos t + (\sin t)e_{13}e_{14}\big)\big(\cos t + (\sin t)e_{15}e_{16}\big)$ and for $t = \pi/2, \pi, 3\pi/2$ this image goes through $e_{11}e_{12}\dots e_{16}$, -1, and $-e_{11}e_{12}\dots e_{16}$ respectively, corresponding to the points i, -1 and $-i$ of S^1 in $\mathbf{C}$.) Thus the subgroup of E_6 is $Spin(10) \times S^1/\mathbf{Z}_4$ as claimed. This completes the check of subgroups mentioned in Theorem 6.1.

Identification of $L(G)$ and $L(G)/L(H)$.

Our tactics are as before. We know $L(E_8) = L(Spin(16)) + \Delta^+$ as a representation of $Spin(16)$ and we have a subgroup $H \times K$ mapping into $Spin(16)$. We wish to determine the centraliser G of K in E_8, so we write $L(E_8)$ as a representation of $H \times K$, where K is one of the groups $SU(2) \subset SU(3) \subset G_2$, and take the part fixed under K. This is $L(G)$ and we regard it as a representation of H.

Case 1. $G = F_4$. By restricting from $Spin(16)$ to $Spin(9) \times Spin(7)$ we have $L(Spin(16)) \mapsto L(Spin(9)) + L(Spin(7)) + \lambda_9^1 \otimes \lambda_7^1$ and $\Delta^+ \mapsto \Delta \otimes \Delta$.

By restriction from $Spin(7)$ to G_2, we have $L(Spin(7)) = L(G_2) + \alpha$ where α is the basic 7-dimensional representation of G_2. (The roots of $Spin(7)$ which are not roots of G_2 are $\pm(x_i + x_j) = \mp x_k$ and 0, once.) Under the restriction, $\lambda_7^1 \mapsto \alpha$, $\Delta \mapsto 1 + \alpha$. Hence

$$L(E_8) = L(Spin(9)) + L(G_2) + \alpha + \lambda_9^1 \otimes \alpha + \Delta \otimes (1 + \alpha).$$

The part fixed by G_2 is $L(Spin(9)) + \Delta \otimes 1$ which is therefore equal to $L(F_4)$.

Note. We now have a map $F_4 \times G_2 \to E_8$, so we may ask: what is $L(E_8)$ as a representation of $F_4 \times G_2$? Well,

$$L(E_8) = L(F_4 \times G_2) + (1 + \lambda_9^1 + \Delta) \otimes \alpha$$

so we have

Corollary 8.1. *There is a representation of F_4 of degree $26 = 1 + 9 + 16$ whose restriction to $Spin(9)$ is $1 + \lambda_9^1 + \Delta$. This representation is real (for compact forms of the groups) and its weights are 0 (twice), $\pm x_i$, $i = 1, 2, 3, 4$, $\frac{1}{2}(\pm x_1 \pm x_2 \pm x_3 \pm x_4)$.* □

Case 2. $G = E_7$. Here we restrict from $Spin(16)$ to $Spin(12) \times Spin(4)$ and $L(Spin(16)) \mapsto L(Spin(12)) + L(Spin(4)) + \lambda_{12} \otimes \lambda_4$ and the representation Δ^+ of $Spin(16)$ restricts to the representation $\Delta^+ \otimes \Delta^+ + \Delta^- \otimes \Delta^-$ of $Spin(12) \times Spin(4)$. Now, $Spin(4) \cong S^3 \times S^3$, where we regard the second factor as $SU(2) = \ker \Delta^+$, and under the isomorphism λ_4^1, Δ^+, Δ^- correspond respectively to $\lambda^1 \otimes \lambda^1$, $\lambda^1 \otimes 1$, $1 \otimes \lambda^1$. Thus

$$L(E_8) = L(Spin(12)) + L(S^3) + L(S^3) + \lambda_{12}^1 \otimes \lambda^1 \otimes \lambda^1 + \Delta^+ \otimes \lambda^1 \otimes 1 + \Delta^- \otimes 1 \otimes \lambda^1.$$

The part fixed under $SU(2)$ is $L(E_7) = L(Spin(12)) + L(S^3) + \Delta^+ \otimes \lambda^1$.

We now have a map $E_7 \times S^3 \to E_8$ and as a representation of $E_7 \times S^3$,

$$L(E_8) = L(E_7 \times S^3) + (\lambda_{12}^1 \otimes \lambda^1 + \Delta^- \otimes 1) \otimes \lambda^1,$$

so we obtain

Corollary 8.2. *E_7 has a representation whose restriction to $Spin(12) \times S^3$ is $\lambda_{12}^1 \otimes \lambda^1 + \Delta^- \otimes 1$, with degree 56, and it is symplectic for compact forms of E_7.* □

Case 3. $G = E_6$. When we restrict from $Spin(16)$ to $Spin(10) \times Spin(6)$, $L(Spin(16))$ restricts to $L(Spin(10)) + L(Spin(6)) + \lambda_{10}^1 \otimes \lambda_6^1$ and Δ^+ restricts to $\Delta^+ \otimes \Delta^+ + \Delta^- \otimes \Delta^-$.

Write ξ for the identity representation of $S^1 = U(1)$. Then, on restricting $Spin(6)$ to $S^1 \times SU(3)$ under our map $S^1 \times SU(3) \to Spin(6)$, we find that $L(Spin(6))$ restricts to $L(S^1 \times SU(3)) + \xi^4 \otimes \lambda^2 + \xi^{-4} \otimes \lambda^1$. (The roots of $Spin(6)$ which are not roots of $S^1 \times SU(3)$ are $\pm(x_i + x_j)$.) By looking at weights, we see that Δ^+, Δ^-, λ_6^1 restrict respectively to $\xi^3 \otimes 1 + \xi^{-1} \otimes \lambda^1$, $\xi^{-3} \otimes 1 + \xi \otimes \lambda^2$, $\xi^2 \otimes \lambda^1 + \xi^{-2} \otimes \lambda^2$. Putting all this together gives

$$\begin{aligned} L(E_8) = {} & L(Spin(10)) + L(S^1) + L(SU(3)) \\ & + \xi^4 \otimes \lambda^2 + \xi^{-4} \otimes \lambda^1 + \lambda_{10}^1 \otimes \xi^2 \otimes \lambda^1 + \lambda_{10}^1 \otimes \xi^{-2} \otimes \lambda^2 \\ & + \Delta^+ \otimes \xi^3 \otimes 1 + \Delta^+ \otimes \xi^{-1} \otimes \lambda^1 + \Delta^- \otimes \xi^{-3} \otimes 1 + \Delta^- \otimes \xi \otimes \lambda^2, \end{aligned}$$

where $L(E_6)$ is the part on which $SU(3)$ acts trivially:

$$L(E_6) = L(Spin(10) \times S^1) + (\Delta^+ \otimes \xi^3 + \Delta^- \otimes \xi^{-3}).$$

Note. We have a map $E_6 \times SU(3) \to E_8$ now, so we may consider $L(E_8)$ as a representation of $E_6 \times SU(3)$. Regarded thus,

$$L(E_8) = L(E_6 \times SU(3)) + (\xi^{-4} + \lambda^1_{10} \otimes \xi^2 + \Delta^+ \otimes \xi^{-1}) \otimes \lambda^1 + (\xi^4 + \lambda^1_{10} \otimes \xi^{-2} + \Delta^- \otimes \xi) \otimes \lambda^2.$$

This leads to

Corollary 8.3. *E_6 has two representations whose restrictions to $Spin(10) \times S^1$ are respectively $\xi^{-4} + \lambda^1_{10} \otimes \xi^2 + \Delta^+ \otimes \xi^{-1}$ and $\xi^4 + \lambda^1_{10} \otimes \xi^{-2} + \Delta^- \otimes \xi$. These are of degree 27 and complex conjugates of each other.* □

This completes the discussion of Theorem 6.1. For further information on representations of exceptional groups, see [Ad 2]. □

The Real Forms of E_8, continued (from Chapter 7).

Our data are: a complex matrix group G; an element $x \in G$ of order 2; $A \subset L(G)$, a Lie subalgebra over $\mathbf{R}$ such that $\mathbf{C} \otimes A \cong L(G)$ and the adjoint action of x preserves A.

Example. $G =$ the complex form of E_8; $x = -1 \in Spin(16) \to E_8$; $A = L(Spin(16)) + \Delta^+$, the Lie algebra for the compact form of E_8. □

We decompose A as $A^+ + A^-$, where x acts as 1 on A^+, and as -1 on A^-.

Example. $A^+ = L(Spin(16))$, $A^- = \Delta^+$. □

To continue, $\mathbf{C} \otimes A^+$ is the Lie algebra of the centraliser of $x \in G$ and $[A^+, A^+] \subset A^+$, $[A^+, A^-] \subset A^-$, $[A^-, A^-] \subset A^+$. In $L(G)$, form $A^+ + iA^-$. (For example, form $L(Spin(16)) + i\Delta^+$). Because $[A^+, iA^-] \subset iA^-$, $[iA^-, iA^-] \subset -A^+ = A^+$, we see that $A^+ + iA^-$ is a Lie subalgebra over $\mathbf{R}$ of $L(G)$ with complexification $L(G)$. (For example, $A^+ + iA^- = L(Spin(16)) + i\Delta^+$.) If the Killing form is non-singular on $L(G)$, there is an easy construction for closed subgroups $H \subset G$ with $L(H) = A^+ + iA^-$, namely

$$H = \{ g \in G \mid g \text{ preserves } A^+ + iA^- \subset L(G) \}.$$

Example. For x, take the image of $-1 \in SU(2) \to Spin(4) \to Spin(12) \times Spin(4) \to Spin(16) \to E_8$. This gives us an element of order 2 in G, the complex form of E_8; $A = L(\text{compact form of } E_8)$. Now x centralises E_7, because the whole of $SU(2)$ does so, and centralises $SU(2)$. Thus $L(E_7 \times SU(2)) \subset A^+$ whereas

$L(E_8)/L(E_7 \times SU(2)) =$ (the 56-dimensional representation of E_7) $\otimes \lambda^1 \subset A^-$, where λ^1 is the representation of $SU(2)$. When we form our new Lie algebra $A^+ + iA^-$ and the new Lie group H over $\mathbf{R}$, the Killing form is negative definite on A^+ of dimension 136 and positive definite on iA^- of dimension 112. Thus the Killing form has signature -24 ($\neq -248, 8$). Therefore we have a third form of E_8. (In fact there are only three though we do not prove this here.) $\square$

Note. $\ker(E_7 \times SU(2) \to E_8(\text{compact})) \cong \mathbf{Z}_2$ generated by $(-e_1e_2\ldots e_{12}, -e_{13}e_{14}e_{15}e_{16})$ where $-e_{13}e_{14}e_{15}e_{16}$ corresponds to $-1 \in SU(2)$. Using the theorem that any conjugacy class meets the maximal torus, to find all possible elements of order 2, we may look in T.

Exercise. Show that $T \subset Spin(16)/\mathbf{Z}_2 \subset E_8$ contains $2^8 - 1 = 255$ elements of order 2, falling into two conjugacy classes (that is, two orbits under the Weyl group of E_8) as follows:

120 in the identity component of the centraliser of $E_7 \times SU(2)/\mathbf{Z}_2$,
135 in the identity component of the centraliser of $Spin(16)/\mathbf{Z}_2$.

(See [Ad 1].)

Chapter 9

The Dynkin diagrams of F_4, E_6, E_7, E_8

The Dynkin diagram and roots of G_2 were given in Theorem 5.5. Using the identification of $L(G)$ in Chapter 8, we complete this task here by giving the roots, simple roots and Dynkin diagrams for the remaining exceptional groups. If two Lie groups are isomorphic, their Lie algebras are isomorphic since $G \mapsto L(G)$ is a functor. Root systems are constructed by choosing a maximal torus for G or a Cartan subalgebra for $L(G)$. Any two maximal tori are conjugate (Corollary 1.7) and any two Cartan subalgebras are conjugate. Construction of a Dynkin diagram requires a further choice: a fundamental Weyl chamber (see the diagrams of the classical groups in Chapter 1, Examples 1.10). The Weyl group acts simply transitively on the Weyl chambers [1, p. 110]. Thus if two (simple) Lie groups are isomorphic, their Dynkin diagrams will be the same. For full details see [Ha, p. 140]. By comparing with the Dynkin diagrams in Chapter 1, we see that no exceptional group is isomorphic to a classical group.

F_4.

Take $T \subset Spin(9) \subset F_4$. Then the roots are

$\pm x_i \pm x_j,\quad 1 \le i < j \le 4$: there are 24 of these, all long;
$\pm x_i,\quad 1 \le i \le 4$: there are 8 of these short roots;

(These 32 roots come from $Spin(9)$.)

$\frac{1}{2}(\pm x_1 \pm x_2 \pm x_3 \pm x_4)$: there are 16 of these short roots, all from Δ.

The Weyl chamber is the one containing (8, 1, 2, 3). All roots take non-zero integer values here and exactly four take the value 1, namely

1 $\frac{1}{2}(x_1 - x_2 - x_3 - x_4)$,
2 x_2,
3 $-x_2 + x_3$,

4 $\quad -x_3 + x_4,$

which must therefore be the simple roots. The Dynkin diagram is

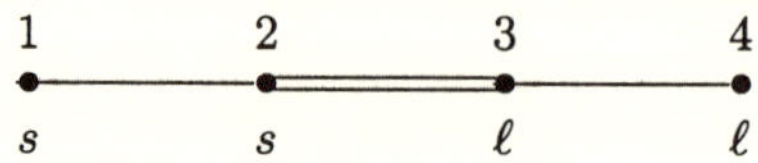

Since the 26-dimensional representation restricts to $1 + \lambda_9^1 + \Delta$ on $Spin(9)$, its weights are 0 (twice) and the 24 short roots.

E_8.

Take $T \subset Spin(16)/\mathbf{Z}_2 \subset E_8$. The roots are

$\pm x_i \pm x_j$: these are from $Spin(16)$ and there are 112 of them;

$\frac{1}{2}(\pm x_1 \pm x_2 \cdots \pm x_8)$, where there are an even number of $-$ signs.

(There are 128 of these from Δ^+.)

All roots have the same length. The Weyl chamber is taken to contain (23, 0, 1, 2, 3, 4, 5, 6). At this point each root takes non-zero integer values and exactly 8 of them are 1, so these are the simple roots:

1 $\quad \frac{1}{2}(x_1 + x_2 - x_3 - x_4 - x_5 - x_6 - x_7 - x_8),$
2 $\quad -x_2 + x_3,$
3 $\quad x_2 + x_3,$
4 $\quad -x_3 + x_4,$
5 $\quad -x_4 + x_5,$
6 $\quad -x_5 + x_6,$
7 $\quad -x_6 + x_7,$
8 $\quad -x_7 + x_8.$

The Dynkin diagram is

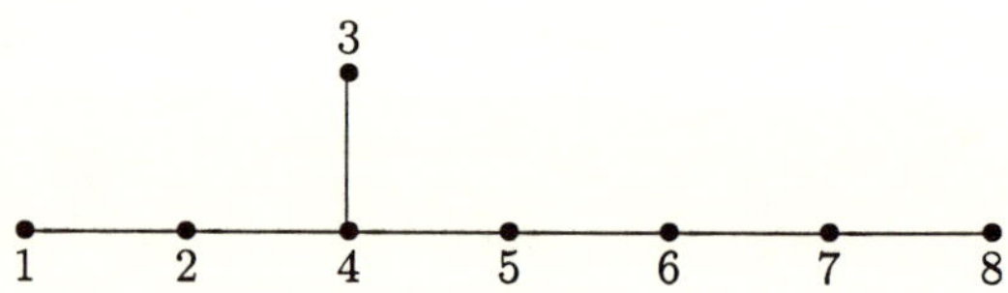

$E_7 \times SU(2)/\mathbf{Z}_2$.

The roots of E_7 are those of E_8 with the same coefficient of x_7 and x_8. Thus the roots are

$\pm(x_7 - x_8)$	(2)	for $SU(2)$,
$\pm(x_7 + x_8)$	(2)	
$\pm x_i \pm x_j \quad 1 \le i < j \le 6$	(60)	
$\frac{1}{2}(\pm x_1 \pm x_2 \cdots \pm x_6 \pm (x_7 + x_8))$	(64)	(where half the signs are $-$).

The Weyl chamber is to contain (13, 0, 1, 2, 3, 4, 0, 1). Hence the simple roots are

1 $\frac{1}{2}(x_1 + x_2 - x_3 - x_4 - x_5 - x_6 - x_7 - x_8)$,
2 $-x_2 + x_3$,
3 $x_2 + x_3$,
4 $-x_3 + x_4$,
5 $-x_4 + x_5$,
6 $-x_5 + x_6$,
7 $-x_7 + x_8$,
8 $x_7 + x_8$.

The Dynkin diagram (of $E_7 \times SU(2)/\mathbf{Z}_2$) is

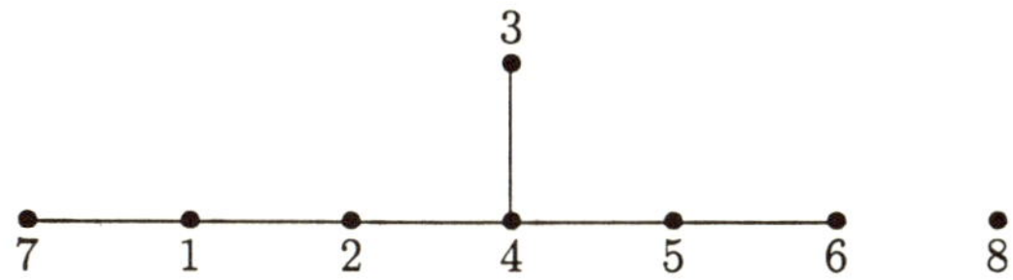

$E_6 \times SU(3)$.

The roots of E_6 are those of E_8 which contain x_6, x_7, x_8 with the same coefficient. Thus the roots are

$\pm(x_i - x_j)$ $\quad 6 \le i < j \le 8$ $\quad$ (6) for $SU(3)$
$\pm x_i \pm x_j$ $\quad 1 \le i < j \le 5$ $\quad$ (40)
$\frac{1}{2}(\pm x_1 \pm x_2 \pm x_3 \pm x_4 \pm x_5 \pm (x_6 + x_7 + x_8))$ $\quad$ (32)
(where we take half of these with $-$ signs).

The Weyl chamber contains $(8, 0, 1, 2, 3, -1, 0, 1)$ and the simple roots are

1 $\frac{1}{2}(x_1 + x_2 - x_3 - x_4 - x_5 - x_6 - x_7 - x_8)$,
2 $-x_2 + x_3$,
3 $x_2 + x_3$,
4 $-x_3 + x_4$,
5 $-x_4 + x_5$,
6 $\frac{1}{2}(x_1 - x_2 - x_3 - x_4 - x_5 + x_6 + x_7 + x_8)$,
7 $-x_6 + x_7$,
8 $-x_7 + x_8$.

The Dynkin diagram (of $E_6 \times SU(3)$) is

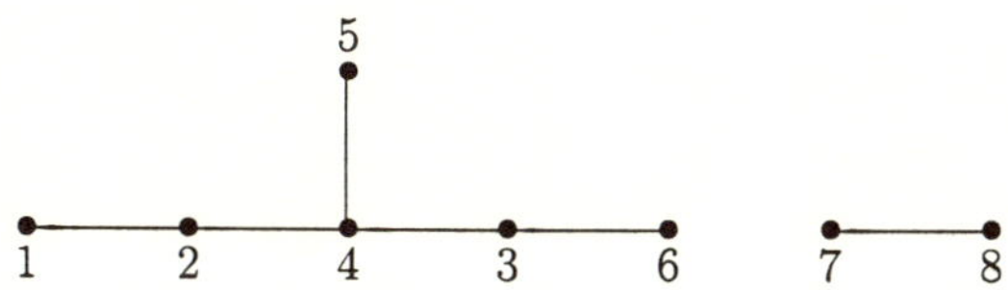

Chapter 10

The Weyl group of E_8

Theorem 10.1. *The Weyl group $W(E_8)$ of E_8 coincides with the group of all automorphisms of the root system of E_8 and its order is $2^{14} \cdot 3^5 \cdot 5^2 \cdot 7$.*

Although we will not need to quote this theorem in the sequel, we will need to rely on several matters introduced in the course of the proof. The proof will involve a series of lemmas and to begin, consider the diagram

$$\begin{array}{ccccc} & & Spin(16) & \longrightarrow & E_8 \\ & \nearrow & \downarrow & & \\ SU(8) \longrightarrow & U(8) \longrightarrow & SO(16). & & \end{array}$$

Lemma 10.2. *The identity component of the centraliser of $SU(8)$ is S^3 (for the compact version of E_8).*

Proof. Consider the action of $SU(8)$ on $L(E_8)$. As an $SU(8)$-module,

$$L(Spin(16)) = L(U(8)) + \lambda^2 + \lambda^6 = L(SU(8)) + 1 + \lambda^2 + \lambda^6,$$
$$\Delta^+ = 1 + \lambda^2 + \lambda^4 + \lambda^6 + 1.$$

Thus $L(\text{centraliser of } SU(8))$ is of dimension 3 and if we work over $\mathbf{C}$, it is generated by $u = e_1e_2 + e_3e_4 + \cdots + e_{15}e_{16}$, v and w, where $[u,v] = 8iv$, $[u,w] = -8iw$, $[v,w] = iu$, see Lemma 6.2. We now have a compact, connected Lie group of rank 1, of dimension 3 and wish to know whether it is $SO(3)$ or S^3. Look at the 1-parameter subgroup with tangent vector u : $\gamma_t = \prod_{r=1}^{8}\Big(\cos t + (\sin t)e_{2r-1}e_{2r}\Big)$. At $t = \pi/2$, this passes through $e_1e_2 \ldots e_{16} = 1$. Change scale to $\frac{1}{4}u = e_1e_2 + \cdots + e_{15}e_{16}$. Then $[\frac{1}{4}u, v] = 2iv$, $[\frac{1}{4}u, w] = -2iw$, $[v,w] = iu$, so the roots are $\pm 2x_1$ and the centraliser is $Sp(1) \cong S^3$. $\square$

Lemma 10.3. *In $W(E_8)$ there exists an element which reverses the vector $u = e_1e_2 + \cdots + e_{15}e_{16} \in L(T)$ and fixes $T(SU(8))$.*

Proof. In S^3, conjugation by j inverts the circle e^{it} and fixes the whole of $SU(8)$ since our S^3 arises as the centraliser of $SU(8)$. □

Lemma 10.4. $W(E_8)$ is transitive on the roots.

Proof. $W(Spin(16)/\mathbf{Z}_2)$ is transitive on the 112 roots, $\pm x_i \pm x_j$, of $Spin(16)$ and also transitive on the 128 weights $\frac{1}{2}(\pm x_1 \pm x_2 \pm \cdots \pm x_8)$ of Δ^+. Thus the roots of E_8 fall into at most two orbits. But the reflection in the hyperplane perpendicular to $u = e_1e_2 + \cdots + e_{15}e_{16}$, given by $\rho_u(v) = v - 2\frac{u \cdot v}{u \cdot u}u$, carries $x_1 + x_2$ to $\frac{1}{2}(x_1 + x_2 - x_3 - \cdots - x_8)$, so there is only one orbit. □

Lemma 10.5. *For each root θ of E_8, E_8 contains a subgroup of local type $T^7 \times S^3$ which contains $T(E_8)$ and whose roots are $\pm\theta$. The map $S^3 \to E_8$ is mono; this subgroup and its S^3 subgroup are uniquely determined by θ and subgroups corresponding to different choices of θ are conjugate. Conjugation with a suitable element of this subgroup gives an element of $W(E_8)$ interchanging θ, $-\theta$ and fixing all weights perpendicular to $\pm\theta$.*

Proof. We already know most of these parts for two pairs of roots, namely for $\pm\frac{1}{2}(x_1+\cdots+x_8)$ by Lemma 10.2 and Lemma 10.3 and for $\pm(x_7-x_8)$ because we have studied the subgroup $SU(2) \to Spin(4) \xrightarrow[\text{coordinates}]{\text{last}} Spin(16) \to E_8$, in Chapter 8 in connection with E_7, and can adjoin the remaining elements of the torus to that. By Lemma 10.4 the result holds for all roots θ. For the uniqueness, look at the 2-dimensional subspace of the real vector space $L(E_8)$ on which T acts via $\pm\theta$ (after complexification). This subspace is unique and under $[\ ,\ ]$ generates a 3-dimensional subalgebra characterising S^3. □

By $\overset{2}{\bullet}\!\!-\!\!\!-\!\!\!-\!\!\overset{1}{\bullet}$ we will mean roots θ_1 and θ_2 of E_8 meeting at $120°$.

Lemma 10.6. *$W(E_8)$ is transitive on the pairs $\overset{2}{\bullet}\!\!-\!\!\!-\!\!\!-\!\!\overset{1}{\bullet}$ and their number is $56 \cdot 240 = 2^7 \cdot 3 \cdot 5 \cdot 7$.*

Proof. By Lemma 10.5, we can take $\theta_1 = -x_7 + x_8$ and count how many roots have inner product -1 with θ_1. We have

$\pm x_i + x_7$,	$i \leq 6$	(12),
$\pm x_i - x_8$,	$i \leq 6$	(12),
$\frac{1}{2}(\pm x_1 \pm \cdots \pm x_6 + x_7 - x_8)$,	even number of $-$ signs	(32),

making a total of 56. There are 240 roots of E_8 (see Chapter 9), so 240 possibilities for θ_1. Using the Weyl group of $Spin(12)$, which fixes θ_1, we may permute $x_1, \ldots, x_6$ in any way and change an even number of signs; in this way we get three orbits. If we use $W(Spin(12) \times Sp(1))$, we have an operation fixing $x_1, \ldots, x_6$ and sending $x_7 \mapsto -x_8$, $x_8 \mapsto -x_7$. This interchanges the first two orbits, so we are down to two orbits of size 24 and 32. Finally the reflection reversing $\frac{1}{2}(x_1+x_2+\cdots+x_8)$ fixes θ_1 and sends x_1+x_7 to $\frac{1}{2}(x_1-x_2-\cdots-x_6+x_7-x_8)$. Hence under $W(E_8)$ there is only one orbit. □

Corollary. *For each pair* $\overset{2}{\bullet}\!\!-\!\!\!-\!\!\overset{1}{\bullet}$, *$E_8$ contains a unique subgroup of local type $T^6 \times SU(3)$ containing $T(E_8)$ and with* $\overset{2}{\bullet}\!\!-\!\!\!-\!\!\overset{1}{\bullet}$ *as its Dynkin diagram. That is, θ_1 and θ_2 qualify as a pair of simple roots for the subgroup. The subgroups arising in this way are all conjugate and $SU(3) \to E_8$ is mono.*

Proof. By transitivity, we need concern ourselves only with $\overset{x_7-x_6}{\bullet}\!\!-\!\!\!-\!\!\!-\!\!\overset{x_8-x_7}{\bullet}$: take the diagram $SU(3) \to Spin(6) \to Spin(16) \to E_8$ and fill out to a maximal torus. □

By a "triple $\overset{3}{\bullet}\!\!-\!\!\!-\!\!\overset{2}{\bullet}\!\!-\!\!\!-\!\!\overset{1}{\bullet}$" we will mean a triple consisting of three roots θ_1, θ_2, θ_3 of E_8 such that θ_1, θ_2 make an angle of 120° and similarly for θ_2, θ_3, while θ_1, θ_3 are orthogonal. Below we use the same language for other configurations of roots.

Lemma 10.7. *$W(E_8)$ is transitive on the triples* $\overset{3}{\bullet}\!\!-\!\!\!-\!\!\overset{2}{\bullet}\!\!-\!\!\!-\!\!\overset{1}{\bullet}$ *and their number is* $2^7 \cdot 3^4 \cdot 5 \cdot 7$.

Proof. By Lemma 10.6, we can take $\overset{2}{\bullet}\!\!-\!\!\!-\!\!\overset{1}{\bullet}$ to be $\overset{x_7-x_6}{\bullet}\!\!-\!\!\!-\!\!\!-\!\!\overset{x_8-x_7}{\bullet}$. Now the roots orthogonal to $-x_7+x_8$ are

$$\begin{array}{ll} \pm(x_7+x_8), & \\ \pm x_i \pm x_j, & 1 \le i < j \le 6, \\ \frac{1}{2}(\pm x_1 \pm \cdots \pm x_6 \pm (x_7+x_8)), & \text{even number of } - \text{ signs,} \end{array}$$

and those which in addition have inner product -1 with $-x_6+x_7$ are

$$\begin{array}{ll} -x_7-x_8, & (1), \\ \pm x_i + x_6, & (10), \\ \frac{1}{2}(\pm x_1 \pm \cdots \pm x_5 - x_6 - x_7 - x_8), & (16), \end{array}$$

making 27 in all. Using the Weyl group of $Spin(10)$, which fixes θ_1, θ_2, we can permute $x_1, \ldots, x_5$ and change an even number of their signs; in this way we get three orbits. The reflection which reverses $\frac{1}{2}(x_1+\cdots+x_8)$ and fixes θ_1,

θ_2, sends $x_1 + x_6$ to $\frac{1}{2}(x_1 - x_2 - \cdots - x_5 + x_6 - x_7 - x_8)$ and $-x_7 - x_8$ to $\frac{1}{2}(x_1 + \cdots + x_6 - x_7 - x_8)$. So under $W(E_8)$ we get only one orbit. $\square$

Corollary. *Each triple* $\overset{3}{\bullet}\!\!-\!\!-\!\!\overset{2}{\bullet}\!\!-\!\!-\!\!\overset{1}{\bullet}$ *arises as the Dynkin diagram of a subgroup of local type* $T^5 \times SU(4)$ *containing* $T(E_8)$.

Proof. The proof is as above. $\square$

Lemma 10.8. $W(E_8)$ *is transitive on the quadruples* $\overset{4}{\bullet}\!\!-\!\!-\!\!\overset{3}{\bullet}\!\!-\!\!-\!\!\overset{2}{\bullet}\!\!-\!\!-\!\!\overset{1}{\bullet}$ *and their number is* $2^{11} \cdot 3^4 \cdot 5 \cdot 7$.

Proof. By the previous work, we may take $\theta_3, \theta_2, \theta_1$ as $-x_5+x_6, -x_6+x_7, -x_7+x_8$ respectively. The roots which are orthogonal to $-x_6+x_7$, $-x_7+x_8$ are $\pm x_i \pm x_j$ with $i < j \leq 5$ and $\frac{1}{2}(\pm x_1 \pm \cdots \pm x_5 \pm (x_6+x_7+x_8))$, (even number of $-$ signs). Of these, the ones which have inner product -1 with $-x_5+x_6$ are $\pm x_i + x_5$, $i \leq 4$ and $\frac{1}{2}(\pm x_1 \pm \cdots \pm x_4 + x_5 - x_6 - x_7 - x_8)$ (even number of $-$ signs), making 16 in all. Using the Weyl group of $Spin(8)$, which fixes θ_1, θ_2, θ_3, we can permute $x_1, \ldots, x_4$ and change an even number of signs. In this way we get two orbits. The reflection which reverses $\frac{1}{2}(x_1 + \cdots + x_8)$ fixes θ_1, θ_2, θ_3 and sends $x_1 + x_5$ to $\frac{1}{2}(x_1 - x_2 - x_3 - x_4 + x_5 - x_6 - x_7 - x_8)$, so we have only one orbit. $\square$

Corollary. *Each quadruple* $\overset{4}{\bullet}\!\!-\!\!-\!\!\overset{3}{\bullet}\!\!-\!\!-\!\!\overset{2}{\bullet}\!\!-\!\!-\!\!\overset{1}{\bullet}$ *arises as the Dynkin diagram of a subgroup of local type* $T^4 \times SU(5)$ *containing* $T(E_8)$.

Proof. As before. $\square$

Lemma 10.9. $W(E_8)$ *is transitive on the 5-tuples* $\overset{5}{\bullet}\!\!-\!\!-\!\!\overset{4}{\bullet}\!\!-\!\!-\!\!\overset{3}{\bullet}\!\!-\!\!-\!\!\overset{2}{\bullet}\!\!-\!\!-\!\!\overset{1}{\bullet}$ *and their number is* $2^{12} \cdot 3^4 \cdot 5^2 \cdot 7$.

Proof. By the previous work, we may take $\theta_4, \ldots, \theta_1$ as $-x_4+x_5, -x_5+x_6, -x_6+x_7, -x_7+x_8$. The roots which are orthogonal to $-x_5+x_6, -x_6+x_7, -x_7+x_8$ are

$$\pm x_i \pm x_j, \qquad i < j \leq 4,$$
$$\tfrac{1}{2}(\pm x_1 \pm x_2 \pm x_3 \pm x_4 \pm (x_5 + x_6 + x_7 + x_8)) \qquad \text{(even number of } - \text{ signs)}.$$

Of these, the ones which have inner product -1 with $-x_4 + x_5$ are

$$\pm x_i + x_4,$$
$$\tfrac{1}{2}(\pm x_1 \pm x_2 \pm x_3 + x_4 - x_5 - x_6 - x_7 - x_8), \qquad \text{(ten in all)}.$$

Using the Weyl group of $Spin(6)$, which fixes $\theta_1, \ldots, \theta_4$, we get two orbits. The reflection reversing $\frac{1}{2}(x_1 + \cdots + x_8)$ fixes $\theta_1, \ldots, \theta_4$ and sends $x_1 + x_4$ to $\frac{1}{2}(x_1 - x_2 - x_3 + x_4 - x_5 - x_6 - x_7 - x_8)$, so we have only one orbit. □

Corollary. *Each 5-tuple* 

arises as the Dynkin diagram of a subgroup of local type $T^3 \times SU(6)$ containing $T(E_8)$. □

Lemma 10.10. *$W(E_8)$ is transitive on the 6-tuples*

and their number is $2^{13} \cdot 3^5 \cdot 5^2 \cdot 7$.

Proof. As above. □

Corollary. *Each 6-tuple* 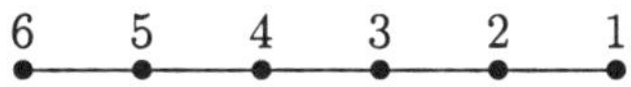

arises as the Dynkin diagram of a subgroup of local type $T^2 \times SU(7)$ containing $T(E_8)$. □

Lemma 10.11. *$W(E_8)$ is transitive on the 7-tuples*

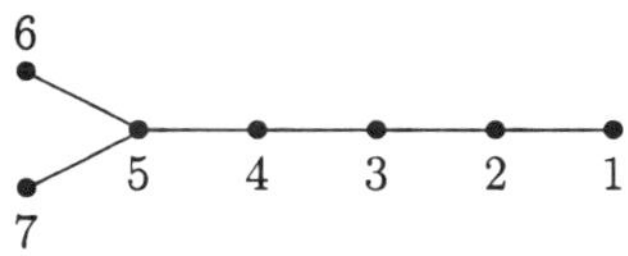

and of these there are $2^{14} \cdot 3^5 \cdot 5^2 \cdot 7$.

Proof. As before. □

Corollary. *Each figure*  *arises as the Dynkin diagram of a subgroup of local type $T^1 \times Spin(14)$.* □

Lemma 10.12. *$W(E_8)$ is transitive on the 8-tuples*

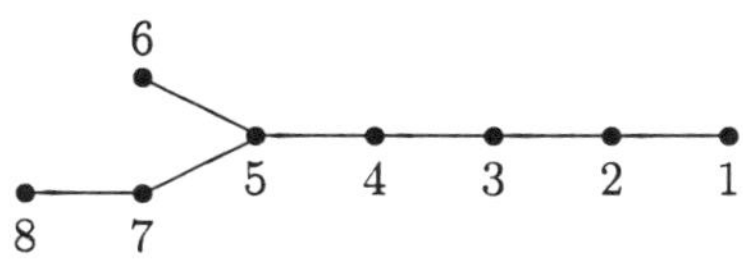

and there are $2^{14} \cdot 3^5 \cdot 5^2 \cdot 7$ of them.

Proof. We may take $\theta_1, \ldots, \theta_7$ to be $-x_7 + x_8$, $-x_6 + x_7$, $\ldots$, $x_2 + x_3$. The roots which are orthogonal to $\theta_1, \ldots, \theta_6$ are $\pm\frac{1}{2}(x_1 + \cdots + x_8)$. The one which has inner product -1 with x_2+x_3 is $-\frac{1}{2}(x_1+\cdots+x_8)$, so each 7-tuple extends to a unique 8-tuple. □

Proof of Theorem 10.1. Since $W(E_8)$ is transitive on the 8-tuples, it follows that the group of all symmetries of the roots system is also transitive. The effect of $W(E_8)$ is determined by its action on these figures as the 8-tuple is a basis for $L(T)^*$ and so both groups are simply transitive on the 8-tuples. Theorem 10.1 follows. □

Now consider unordered figures of the form

,

where we have 8 roots of E_8 which fall into four pairs in orthogonal 2-spaces, each making an angle of 120°.

Theorem 10.13. *Such figures are permuted transitively by* $W(E_8)$. *Each arises as the Dynkin diagram of a subgroup* $H \supset T(E_8)$ *where* $H = (SU(3) \times SU(3) \times SU(3) \times SU(3))/\mathbf{Z}_3 \times \mathbf{Z}_3$ *and* $Z(H) = \mathbf{Z}_3 \times \mathbf{Z}_3$. *From two non-trivial elements in* $Z(SU(3)) \cong \mathbf{Z}_3$ *(four times) we obtain the* 8 *non-trivial elements in* $Z(H)$. *Let* $S \subset W(E_8)$ *be the subgroup preserving a figure. Then* $S \to Aut(Z(H)) = GL(2, \mathbf{F}_3)$ *is an isomorphism. In particular,* S *maps onto* Σ_4, *the permutations of the four* $SU(3)$ *and the kernel of* $S \to \Sigma_4 = PGL(2, \mathbf{F}_3)$ *(corresponding to* $-1 \in GL(2, \mathbf{F}_3)$) *acts as* 

Explanation. Here $Z(G)$ is the centre of a group G. $Out(SU(3)) \cong \mathbf{Z}_2$ acts by inverting both the centre of $SU(3)$ and the Dynkin diagram. In other words, the point where one root of $SU(3)$ is 1 and the other -1 is a non-trivial element of the centre, so if you turn the Dynkin diagram end to end, you invert the centre and vice versa.

Proof. This will take up the rest of the chapter. We continue in the same spirit as before.

Lemma 10.14. $W(E_8)$ *is transitive on the figures*

and there are $2^{10} \cdot 3^3 \cdot 5 \cdot 7$ *of these.*

Proof. By the proof of Lemma 10.12, we may take θ_1 and θ_2 as 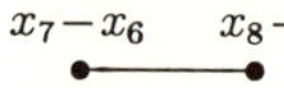

. Orthogonal to both we have

$$\pm x_i \pm x_j, \qquad 1 \le i < j \le 5 \tag{40}$$

$$\tfrac{1}{2}(\pm x_1 \pm \cdots \pm x_5 \pm (x_6 + x_7 + x_8)), \qquad \text{even number of } - \text{ signs} \quad (32),$$

making 72 roots.

Under $W(Spin(10))$, which fixes θ_1, θ_2, we have orbits

$$\begin{aligned}&\pm x_i \pm x_j,\\ &\tfrac{1}{2}(\pm x_1 \pm \cdots \pm x_5 + (x_6 + x_7 + x_8)),\\ &\tfrac{1}{2}(\pm x_1 \pm \cdots \pm x_5 - (x_6 + x_7 + x_8)).\end{aligned}$$

Using the reflection which reverses $\frac{1}{2}(x_1 + \cdots + x_8)$ and fixes θ_1, θ_2, we send $x_1 + x_2$ to $\frac{1}{2}(x_1 + x_2 - x_3 - \cdots - x_8)$. Hence there is only one orbit under $W(E_8)$. ∎

Lemma 10.15. $W(E_8)$ *is transitive on the figures*

and there are $2^{12} \cdot 3^3 \cdot 5^2 \cdot 7$ *of these.*

Proof. By the previous lemma we may take θ_3, θ_2, θ_1 as

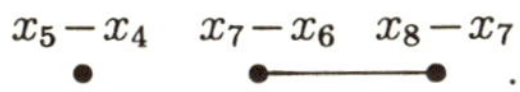

The roots orthogonal to θ_1, θ_2 and of inner product -1 with θ_3 are

$$\begin{aligned}&\pm x_i + x_4, && 1 \le i \le 3 \quad (6),\\ &\pm x_i - x_5, && 1 \le i \le 3 \quad (6),\\ &\tfrac{1}{2}(\pm x_1 \pm x_2 \pm x_3 + x_4 - x_5 \pm (x_6 + x_7 + x_8)),\\ & && \text{even number of } - \text{ signs} \quad (8),\end{aligned}$$

making 20 roots.

Under $W(Spin(6))$, which fixes θ_1, θ_2, θ_3, we have orbits

$$\begin{aligned}&\pm x_i + x_4,\\ &\pm x_i - x_5,\\ &\tfrac{1}{2}(\pm x_1 \pm x_2 \pm x_3 - x_4 - x_5 + x_6 + x_7 + x_8),\\ &\tfrac{1}{2}(\pm x_1 \pm x_2 \pm x_3 - x_4 - x_5 - x_6 - x_7 - x_8).\end{aligned}$$

In $W(E_8)$ we have $x_4 \mapsto -x_5$, $x_5 \mapsto -x_4$ which fixes θ_1, θ_2, θ_3 and sends the first two orbits into one orbit of 12 elements. By using the reflection reversing $\frac{1}{2}(x_1 + \cdots + x_8)$ which fixes θ_1, θ_2, θ_3, we send $x_1 + x_4$ to $\frac{1}{2}(x_1 - x_2 - x_3 + x_4 - x_5 - \cdots - x_8)$ and $-x_1 - x_5$ to $\frac{1}{2}(-x_1 + x_2 + x_3 + x_4 - x_5 + x_6 + x_7 + x_8)$. Thus there is only one orbit under $W(E_8)$. ∎

Corollary. *Any figure* 

arises as the Dynkin diagram of a subgroup of local type $T^4 \times SU(3) \times SU(3)$. *Moreover,* $SU(3) \times SU(3) \to E_8$ *is mono.*

Proof. It is sufficient to check for $-x_3 + x_4$, $-x_4 + x_5$, $-x_6 + x_7$, $-x_7 + x_8$ and we can put $SU(3) \times SU(3) \to Spin(6) \times Spin(6) \to Spin(16)/\mathbf{Z}_2 \to E_8$. □

Lemma 10.16. *For each figure* 

the roots orthogonal to $\theta_1, \theta_2, \theta_3, \theta_4$ *form a root system of type* $A_2 \times A_2$.

Proof. By the previous lemma, we need only consider

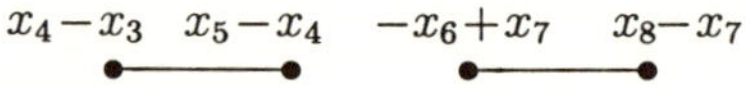

The roots orthogonal to these are

$$\begin{aligned} &\pm x_1 \pm x_2, && (4), \\ &\tfrac{1}{2}(\pm x_1 \pm x_2 \pm (x_3 + x_4 + x_5) \pm (x_6 + x_7 + x_8)), && (8). \end{aligned}$$

These split into the following subsets

$$\begin{aligned} &\pm (x_1 + x_2), \\ &\pm \tfrac{1}{2}(-x_1 - x_2 + x_3 + \cdots + x_8), \\ &\pm \tfrac{1}{2}(-x_1 - x_2 - \cdots - x_8) \end{aligned}$$

and

$$\begin{aligned} &\pm (x_1 - x_2), \\ &\pm \tfrac{1}{2}(-x_1 + x_2 + x_3 + x_4 + x_5 - x_6 - x_7 - x_8), \\ &\pm \tfrac{1}{2}(-x_1 + x_2 - x_3 - x_4 - x_5 + x_6 + x_7 + x_8). \end{aligned}$$

These two subsets have the form $\pm\alpha$, $\pm\beta$, $\pm\gamma$, $\alpha + \beta + \gamma = 0$, (which is the root system for $SU(3)$) and lie in orthogonal 2-planes. □

It is now clear that the number of ways of choosing $\theta_5, \ldots, \theta_8$ is $12 \cdot 2 \cdot 6 \cdot 2 = 2^5 \cdot 3^2$.

Lemma 10.17. $W(E_8)$ *is transitive on the figures*

(*of which there are* $2^{10} \cdot 3^4 \cdot 5^2 \cdot 7$) *and the stabiliser has order* ≥ 48.

Proof. To count the number, we have to divide the number of figures, namely $2^5 \cdot 3^2 \times 2^{12} \cdot 3^3 \cdot 5^2 \cdot 7$, by the number of ways of putting the numbers on, which is $2 \cdot 4 \cdot 6 \cdot 8 = 2^7 \cdot 3$. To prove transitivity, take any such figure and write the two figures concerned as

$$\overset{\theta_8}{\bullet}\!\!-\!\!\!-\!\!\!-\!\!\overset{\theta_7}{\bullet} \quad \overset{\theta_6}{\bullet}\!\!-\!\!\!-\!\!\!-\!\!\overset{\theta_5}{\bullet} \quad \overset{\theta_4}{\bullet}\!\!-\!\!\!-\!\!\!-\!\!\overset{\theta_3}{\bullet} \quad \overset{\theta_2}{\bullet}\!\!-\!\!\!-\!\!\!-\!\!\overset{\theta_1}{\bullet},$$

$$\overset{\phi_8}{\bullet}\!\!-\!\!\!-\!\!\!-\!\!\overset{\phi_7}{\bullet} \quad \overset{\phi_6}{\bullet}\!\!-\!\!\!-\!\!\!-\!\!\overset{\phi_5}{\bullet} \quad \overset{\phi_4}{\bullet}\!\!-\!\!\!-\!\!\!-\!\!\overset{\phi_3}{\bullet} \quad \overset{\phi_2}{\bullet}\!\!-\!\!\!-\!\!\!-\!\!\overset{\phi_1}{\bullet}.$$

Choose a root θ_1 and a root θ_3 (there are 48 possibilities). By Lemma 10.15, there is an element of $W(E_8)$ taking $\theta_1, \ldots, \theta_4$ onto our standard roots $\phi_1, \ldots, \phi_4$. This standard choice of $\phi_1, \ldots, \phi_4$ determines an orthogonal root system of type $A_2 \times A_2$. The reflections corresponding to these roots generate a subgroup $\Sigma_3 \times \Sigma_3$ of $W(E_8)$ fixing $\phi_1, \ldots, \phi_4$. (In fact, by the corollary to Lemma 10.15, this $\Sigma_3 \times \Sigma_3$ actually comes as the Weyl group of a subgroup of local type $T^4 \times SU(3) \times SU(3)$.) Now, θ_5 and θ_6 must go to a pair of roots of one A_2-system, but we do not know which. However, by using the Σ_3 of that A_2-system, we can make sure that θ_5, θ_6 go to our preferred pair of simple roots, but we do not know which one θ_5 goes to or which one θ_6 goes to. Similarly, θ_7 and θ_8 go to a pair of roots of the other A_2-system and by using the Σ_3 of that A_2-system, we can ensure that θ_7, θ_8 go to our preferred pair of simple roots in some order. Thus $W(E_8)$ acts transitively on these figures and we can carry one figure into any other in ≥ 48 ways. □

Lemma 10.18. *Each figure* •——• •——• •——• •——• *arises as the Dynkin diagram of a subgroup of local type* $SU(3)^4$.

Proof. The diagram •——• corresponds to a unique $SU(3)$ subgroup and any two of the four subgroups commute since •——• •——• corresponds to a unique $SU(3) \times SU(3)$. Thus we have a map $SU(3) \times SU(3) \times SU(3) \times SU(3) \to E_8$. □

Continuing with the proof of Theorem 10.13, to show that a subgroup $\mathbf{Z}_3 \times \mathbf{Z}_3$ maps to the identity we can use the coordinates from Lemma 10.16. The centres of the four A_2 work out as follows [1, Proposition 5.3]. We are to solve $\theta_r(t) \equiv 0 \bmod 1$ for $t \in T$, $1 \leq r \leq 8$, and find that the non-zero solutions are given by the following:

$$\pm\tfrac{1}{3}(0,0,1,1,1,0,0,0), \quad \pm\tfrac{1}{3}(0,0,0,0,0,1,1,1),$$
$$\pm\tfrac{1}{3}(0,0,1,1,1,1,1,1), \quad \pm\tfrac{1}{3}(0,0,1,1,1,-1,-1,-1). \quad \text{(All coordinates mod 1.)}$$

The kernel of $SU(3)^4 \to E_8$ must be contained in the centre. The centre of $SU(3)^4$ is $\mathbf{Z}_3 \times \mathbf{Z}_3 \times \mathbf{Z}_3 \times \mathbf{Z}_3$ and at least $\mathbf{Z}_3 \times \mathbf{Z}_3$ of it survives because we have seen that the map $SU(3)^2 \to E_8$ is mono. So the subgroup of E_8 is exactly $SU(3)^4/\mathbf{Z}_3 \times \mathbf{Z}_3$. Each map $SU(3) \to E_8$ is mono, so from the two non-trivial elements in the centre of each $SU(3)$ we get two non-trivial elements in the centre $Z(H)$ and the pairs we get from distinct $SU(3)$ are disjoint as each map $SU(3)^2 \to E_8$ is mono. So, we obtain the 8 non-trivial elements of $Z(H)$.

Lemma 10.19. *Let $S \subset W(E_8)$ be the subgroup preserving a figure. Then there is a monomorphism $S \to Aut(Z(H)) = GL(2, \mathbf{F}_3)$.*

Proof. Suppose $w \in S \subset W(E_8)$ maps to 1. If w preserves the four pairs in $Z(H)$, it must preserve the four $SU(3)$; if it keeps each pair in $Z(H)$ fixed rather than reversing it, it must fix each •——• rather than reversing it, so it fixes $\theta_1, \ldots, \theta_8$. But these form a basis, so it fixes $L(T)^*$ and $w = 1$. □

Lemma 10.20. *$S \to Aut(Z(H))$ is epi.*

Proof 1. By Lemma 10.17, $|S| \geq 48$, so by Lemma 10.19, the image of S has order ≥ 48. But $|GL(2, \mathbf{F}_3)| = 48$, so the image is $GL(2, \mathbf{F}_3)$.

Proof 2. By referring to the proof of Lemma 10.17, we see that S allows us to take any non-zero $a \in Z(H)$ onto any non-zero $a' \in Z(H)$ and similarly any $b \neq 0, \pm a \in Z(H)$ onto $a'' \neq 0, \pm a'$. □

That finishes the proof of Theorem 10.13. □

Chapter 11

Representations of E_6 and E_7

We now deduce some corollaries for the representations of E_7 and E_6 from the work of Chapter 10. In Chapters 12 and 13 these will serve as guidelines for constructing E_7 and E_6 and their representations independently of Chapter 8.

Theorem 11.1. *The group E_7 contains a subgroup $SU(8)/\{\pm 1\}$ and on restriction to this group, $L(E_7) \longmapsto L(SU(8)) + \lambda^4$ and the (56-dimensional representation of E_7) $\longmapsto \lambda^2 + \lambda^6$.*

Proof. E_7 is the identity component of the centraliser of $SU(2) \subset E_8$, corresponding to $\pm(-x_7+x_8)$. We can equally well use $SU(2) \subset E_8$ corresponding to $\pm\frac{1}{2}(x_1+x_2+\cdots+x_8)$ and this $SU(2)$ is the centraliser of $SU(8)$, (Lemma 10.2), so we get $SU(8) \to E_7$. We need to know the kernel of this map and this comes down to seeing if the point $e_1e_2\dots e_{16} \in Spin(16)$ lies in the image of $SU(8)$ or not. We have the diagram

$$\begin{array}{ccc} & & Spin(16) \\ & \nearrow & \downarrow \\ SU(8) & \longrightarrow & SO(16) \end{array}$$

and the 1-parameter subgroup

$$\prod_{r=1}^{4}\Big(\cos t + (\sin t)e_{2r-1}e_{2r}\Big)\prod_{r=5}^{8}\Big(\cos t - (\sin t)e_{2r-1}e_{2r}\Big)$$

is contained in the torus of the lift of $SU(8)$. By taking $t = \pi/2$, we see that $e_1e_2\dots e_{16}$ lies in the lift of $SU(8)$. Since it covers the element $-1 \in SO(16)$, it corresponds to the matrix $-1 \in SU(8)$, so we get $SU(8)/\{\pm 1\} \to E_7$. □

Now take $L(E_8)$ as a representation of $E_7 \times SU(2)$ and restrict further to $SU(8) \times SU(2)$. Then, as an $SU(8) \times SU(2)$-module,

$$L(E_8) = L(SU(8)) + \lambda^4 + 2(\lambda^2 + \lambda^6) + 3$$

where the "3" arises from $L(SU(2))$.

Corollary 11.2. *The 56-dimensional representation, $\lambda^2 + \lambda^6$, of E_7 is irreducible.*

Proof. On restricting to $Spin(12)\times S^3$ we get the two irreducible summands of degree 24 and 32, but on restricting to $SU(8)$ we get two irreducible summands of degree 28. □

Corollary 11.3. *The action of the summand $\lambda^4 \subset L(E_7)$ on both λ^2 and λ^6 in the 56-dimensional representation is non-zero.*

Proof 1. If λ^4 acted as 0 on λ^2, then λ^2 would be a representation of $L(E_7)$; this is a contradiction since the 56-dimensional representation is irreducible. The same argument works for λ^6.

Proof 2. Let v_α, $v_\beta \in L(E_8)$ be non-zero eigenvectors for T corresponding to the roots α, β. Then $[v_\alpha, v_\beta]$ is an eigenvector for T with weight $\alpha+\beta$. Thus $[v_\alpha, v_\beta] = 0$ unless $\alpha+\beta=0$ or is a root of E_8. To complete the proof we use

Lemma. If $\alpha+\beta$ is a root of E_8, then $[v_\alpha, v_\beta] \neq 0$.

Proof. If $\alpha+\beta$ is a root, then the angle between α and β is 120° $\left(\overset{\alpha}{\bullet}\!\!-\!\!\!-\!\!\!-\!\!\overset{\beta}{\bullet}\right)$ so without loss of generality, it is sufficient to check the lemma for a subgroup $SU(3)$. For classical groups the lemma is trivial. If $\alpha+\beta=0$, we use the transitivity of $W(E_8)$ on roots to reduce to the case of $SU(2)$. □

Theorem 11.4. *The group E_6 contains a subgroup isomorphic to $SU(3)\times SU(3)\times SU(3)/\mathbf{Z}_3$ in such a way that*

$$L(E_6) \longmapsto L(SU(3)^3) + \lambda^1(1)\lambda^1(2)\lambda^1(3) + \lambda^2(1)\lambda^2(2)\lambda^2(3).$$

One of the 27-dimensional representations of E_6 maps to

$$\lambda^1(1)\lambda^2(2) + \lambda^1(2)\lambda^2(3) + \lambda^1(3)\lambda^2(1),$$

whilst the other maps to

$$\lambda^2(1)\lambda^1(2) + \lambda^2(2)\lambda^1(3) + \lambda^2(3)\lambda^1(1),$$

where $\lambda(i)$ denotes the usual space on which the i-th copy of $SU(3)$ acts.

Proof. We defined E_6 as the identity component of the centraliser of $SU(3)$ in E_8 and we have maps

$$SU(3)^4 \to E_8, \quad \text{and}$$

$$SU(3)^3 \to E_6, \quad \text{(sending the last to the centralised } SU(3)).$$

The latter has kernel $\mathbf{Z}_3$.

We regard $L(E_8)$ as a representation of $SU(3)^4$. Now, $L(E_8)$ has 240 roots and $L(SU(3)^4)$ has 24. Each of the $240 - 24 = 216$ roots which are not roots of $L(SU(3)^4)$ must restrict to a sum of weights of the four copies of $SU(3)$. But $SU(3)$ has only one sort of weight shorter than a root, namely $\pm x_1$, $\pm x_2$, $\pm x_3$, and the weights of λ^1, λ^2. The length of each of these is $\frac{1}{\sqrt{3}}$ times the length of a root. We get one of these roots in three of the four cases, 0 in the fourth.

One can check this directly or as follows. If a root of E_8 restricts to 0 on two copies of $SU(3)$, then it is orthogonal to •——• •——• , so by Lemma 10.16, it is a root of $SU(3) \times SU(3)$. Thus, apart from the roots of $SU(3)^4$, a root of E_8 restricts non-trivially on at least three copies of $SU(3)$. The roots of E_6 which are not roots of $SU(3)^3$ give us $78 - 24 = 54$ roots of E_8 which are not roots of $SU(3)^4$ but are 0 on the last $SU(3)$ and we have four copies of $SU(3)$, so we get $4 \cdot 54 = 216$ roots which are non-zero on three, zero on one. There are none left to be non-zero on all four copies. We conclude that these 216 roots are the weights of a representation of $SU(3)^4$ which is a sum of terms $\lambda^{\varepsilon_1}(i_1)\lambda^{\varepsilon_2}(i_2)\lambda^{\varepsilon_3}(i_3)$. Thus

$$L(E_8) = L(SU(3)^4) + \text{sum of terms } \lambda^{\varepsilon_1}(i_1)\lambda^{\varepsilon_2}(i_2)\lambda^{\varepsilon_3}(i_3),$$

where $\varepsilon_i = 1$ or 2 and i_1, i_2, i_3 are 3 of 1,2,3,4 and each term is of degree 27. We get two such terms not involving factor 1 (respectively 2,3,4). Choose non-trivial elements z_1, z_2, z_3 in the centres of the first three copies of $SU(3)$ so that $(z_1, z_2, z_3) \mapsto 1 \in E_8$ and choose isomorphisms with the standard $SU(3)$ so that z_1, z_2, z_3 correspond to some element ωI. Then any element of our symmetry group S (see Theorem 10.13) which preserves the fourth $SU(3)$ and permutes the first three $SU(3)$ must either fix or invert $\{z_1, z_2, z_3\}$, that is, must either permute the first three $SU(3)$ or induce the outer automorphism •⇄• on all of them. Thus S must contain a Σ_3 permuting the first three copies of $SU(3)$ preserving their isomorphism with the standard $SU(3)$ while fixing the last $SU(3)$, or reversing it according to the sign of the permutation.

It is now clear that, of the representations $\lambda^{\varepsilon_1}(1)\lambda^{\varepsilon_2}(2)\lambda^{\varepsilon_3}(3)$, the only ones which are trivial on the kernel of $SU(3)^3 \to E_8$ are

$$\lambda^1(1)\lambda^1(2)\lambda^1(3) \quad \text{and} \quad \lambda^2(1)\lambda^2(2)\lambda^2(3).$$

The subgroup of S which permutes the last three $SU(3)$ fixes $\lambda^1(1)\lambda^1(2)\lambda^1(3)$ and its 8 cosets give the 8 representations we want.

In its guise as $GL(2, \mathbf{F}_3)$, S contains $\left(\begin{smallmatrix} 1 & 0 \\ 0 & -1 \end{smallmatrix}\right)$ which fixes one line, reverses one line and interchanges the remaining two. Thus S contains an element of order 2 fixing the first $SU(3)$, reversing the second and interchanging copies three and four. Identify copies three and four under this symmetry. Then we have two terms $\lambda^1(1)\lambda^2(2)\lambda^1(4)$ and $\lambda^2(1)\lambda^1(2)\lambda^2(4)$. Permuting the first three copies of $SU(3)$ gives $\lambda^1(2)\lambda^2(3)\lambda^1(4)$, $\lambda^1(3)\lambda^2(1)\lambda^1(4)$, $\lambda^2(2)\lambda^1(3)\lambda^2(4)$ and $\lambda^2(3)\lambda^1(1)\lambda^2(4)$. The coefficient of $\lambda^1(4)$ is one 27-dimensional representation and the coefficient of $\lambda^2(4)$ the other. □

Corollary 11.5. *The 27-dimensional representations of E_6 are irreducible.*

Proof. On $Spin(10) \times S^1$ there are irreducible summands of degree 1, 10, 16 and on $SU(3)^3$ there are summands of degree 9, 9, 9. □

Corollary 11.6. *The action of $\lambda^1(1)\lambda^1(2)\lambda^1(3) \subset L(E_6)$ on each of the summands $\lambda^1(1)\lambda^1(2), \ldots,$ of each 27-dimensional representation is non-trivial and similarly for $\lambda^2(1)\lambda^2(2)\lambda^2(3)$.*

Proof 1. See Proof 2 for E_7.

Proof 2. (Cf. Proof 1 for E_7, Corollary 11.3.) Suppose that the action of both
$\lambda^1(1)\lambda^1(2)\lambda^1(3)$ and $\lambda^2(1)\lambda^2(2)\lambda^2(3)$ on all the three summands $\lambda^1(1)\lambda^2(2)$, $\lambda^1(2)\lambda^2(3)$, $\lambda^1(3)\lambda^2(1)$ (of one representation) is zero. Then the three summands would be representations of $L(E_6)$, contradiction. Therefore there is a non-zero one, say the action of $\lambda^1(1)\lambda^1(2)\lambda^1(3)$ on $\lambda^1(1)\lambda^2(2)$. Then by cyclic permutation of $\lambda^1(1)\lambda^1(2)\lambda^1(3)$ the action on $\lambda^1(2)\lambda^2(3)$ and $\lambda^1(3)\lambda^2(1)$ is also non-zero. The representations are dual, hence $\lambda^1(1)\lambda^1(2)\lambda^1(3)$ acts non-trivially on one and hence all of $\lambda^2(1)\lambda^1(2), \ldots$. Apply $-1 \in S$ and see that $\lambda^2(1)\lambda^2(2)\lambda^2(3)$ acts non-trivially on all summands. By duality, it acts non-trivially on all summands of the second representation. □

Chapter 12

Direct construction of E_7

The object of this chapter is to give an explicit construction for E_7, its 56-dimensional representation, without assuming Chapter 8 and show that what we construct is indeed isomorphic to what we have already seen. We will exhibit E_7 as a group of linear maps in 56-dimensional space. In Theorem 14.18 we will see that the 26-dimensional representation of F_4 is a subspace of a Jordan algebra. The 56-dimensional representation W of E_7 cannot have an invariant algebra structure for the simple reason that the non-trivial element in the centre of E_7, say $iI \in SU(8)/\{\pm 1\}$, acts as -1 on W, therefore as 1 on $W \otimes W$; therefore any invariant map $f : W \otimes W \to W$ is identically zero. We will reveal all the algebraic structure that seems to be available, but the reader may feel that it is no better than it has to be. (For structures on W, see [All] and [Bro].) The construction will be in terms of exterior algebra.

To begin with, suppose given two dual vector spaces V, V^* over **C**. We wish to preserve the symmetry between V, V^* so we lay down that V^{**} is to be identified with V in such a way that $(v, v^*) = (v^*, v)$. Now introduce the exterior powers $\lambda^i(V)$, $\lambda^j(V^*)$. These are the components of the exterior algebras $\Lambda(V)$ and $\Lambda(V^*)$, so we have product maps $\lambda^i(V) \otimes \lambda^j(V) \to \lambda^{i+j}(V)$ and $\lambda^i(V^*) \otimes \lambda^j(V^*) \to \lambda^{i+j}(V^*)$. The spaces $\lambda^i(V)$ and $\lambda^i(V^*)$ are dual in such a way that

$$(v_1^* v_2^* \dots v_i^*,\ w_1 w_2 \dots w_i) = \sum_{\sigma \in \Sigma_i} \varepsilon(\sigma)(v_1^*, w_{\sigma(1)})(v_2^*, w_{\sigma(2)}) \dots (v_i^*, w_{\sigma(i)})$$

and this formula is self dual. The space $V \otimes V^*$ acts on V via $(u \otimes v^*)w = u(v^*, w)$. This identifies $V \otimes V^*$ with $Hom(V, V) = L(GL(V))$, the Lie algebra of the group $GL(V)$. Dually $V^* \otimes V$ acts on V^* and this identifies $V^* \otimes V$ with $Hom(V^*, V^*) = L(GL(V^*))$, the Lie algebra of the group $GL(V^*)$. We normally identify $GL(V)$ with $GL(V^*)$ by $(gv^*, v) = (v^*, g^{-1}v)$. This has the effect of making the pairing invariant: $(gv^*, gv) = (v^*, g^{-1}gv) = (v^*, v)$. If we want to proceed infinitesimally with the Lie algebra, the statement that the Lie algebra preserves the pairing is $(\gamma v^*, v) + (v^*, \gamma v) = 0$. Accordingly, we identify

$v \otimes w^* \in L(GL(V))$ with $-w^* \otimes v \in V^* \otimes V = L(GL(V^*))$. The identification preserves the Lie bracket and the formula is self dual.

Next we make explicit the action of $L(GL(V))$ on $\lambda^i(V)$. If we proceed globally, then an element $g \in GL(V)$ acts on $\lambda^i(V)$ as follows:

$$g(v_1v_2 \dots v_i) = (gv_1)(gv_2)\dots(gv_i).$$

However, infinitesimally, we want to see the action of the Lie algebra, so we must put $g = 1 + \gamma t + O(t^2)$ and let $t \to 0$, retaining the terms of first order in t. We get

$$\gamma(v_1v_2 \dots v_i) = \sum_{r=1}^{i} v_1v_2 \dots v_{r-1}(\gamma v_r)v_{r+1} \dots v_i.$$

Let the dimension of V be n and choose an isomorphism $\lambda^n(V) \cong \mathbf{C}$, and dually $\lambda^n(V^*) \cong \mathbf{C}$. The symmetry group preserving these isomorphisms is $SL(V)$, identified with $SL(V^*)$ as above. All subsequent constructions will be invariant under this symmetry group. Its Lie algebra $L(SL(V)) \cong L(SL(V^*))$ is the set of matrices of trace 0, which may be identified with the kernel of $(\,,\,) : V \otimes V^* \to \mathbf{C}$ and $(\,,\,) : V^* \otimes V \to \mathbf{C}$.

Consider now the product map $\lambda^i(V) \otimes \lambda^j(V) \to \lambda^n(V) \cong \mathbf{C}$, assuming $i+j = n$. This map establishes $\lambda^i(V)$ and $\lambda^j(V)$ as dual vector spaces. Therefore we identify $\lambda^i(V)$ with $(\lambda^j(V))^*$, which is already identified with $\lambda^j(V^*)$. It is very important to maintain the symmetry between V and V^*: if we apply the same considerations with V replaced by V^*, we see that $\lambda^j(V^*)$ can be identified with $\lambda^i(V^{**}) = \lambda^i(V)$. This identification only agrees with the previous one up to a sign $(-1)^{ij}$. In the applications, ij will be even, so the sign is $+$. To prove the assertion about the sign, choose a basis $e_1, \dots, e_n$ of determinant 1, that is, the isomorphism $\lambda^n(V) \cong \mathbf{C}$ carries $e_1e_2 \dots e_n$ to 1. Let $e_1^*, \dots, e_n^*$ be the dual basis in V^*; then $e_1^*, \dots, e_n^*$ is also a basis of determinant 1 and our situation is symmetrical. In $\lambda^i(V)$ and $\lambda^j(V)$ consider the elements $e_{r_1} \dots e_{r_i}$ and $e_{s_1} \dots e_{s_j}$ where $1 \le r_1 < \dots < r_i \le n$ and $1 \le s_1 < \dots < s_j \le n$. The product $(e_{r_1} \dots e_{r_i})(e_{s_1} \dots e_{s_j})$ is 0 only if any r is equal to an s and there remains a unique choice of $s_1, \dots, s_j$ for which the r's and s's are all distinct. Then the product is $\varepsilon = \pm 1$, where ε is the signature of the permutation $(r_1 \dots r_i s_1 \dots s_j)$. This says that the element $e_{r_1} \dots e_{r_i} \in \lambda^i(V)$ is to be identified with $\varepsilon e_{s_1}^* \dots e_{s_j}^* \in \lambda^j(V^*)$. Replacing V by V^*, this element in turn has to be identified with $\varepsilon\eta e_{r_1} \dots e_{r_i}$ where η is the signature of the permutation $(s_1 \dots s_j r_1 \dots r_i)$. Of course $\varepsilon\eta = (-1)^{ij}$.

Construction of $L(E_7)$ and its 56-dimensional representation.

We are now ready to construct candidates for $L(E_7)$ and its 56-dimensional representation.

Definition. *Put* $A = L \oplus \lambda^4$ *where* $L = \mathbf{C} \otimes L(SU(8)) = L(SL(V)) = L(SL(V^*))$ *and* $\lambda^4 = \lambda^4(V) = \lambda^4(V^*)$ *where* $\dim V = 8$. *Put* $W = \lambda^2 + \lambda^{2*}$, *where* $\lambda^2 = \lambda^2(V) = \lambda^6(V^*)$.

We wish to see A acting as a Lie algebra of linear maps $W \to W$. The components of this action are

$$L \otimes \lambda^2 \to \lambda^2, \quad \text{the usual action of } L = L(SL(V)) \text{ on } \lambda^2(V),$$
$$L \otimes \lambda^{2*} \to \lambda^{2*}, \quad \text{the usual action of } L \text{ on } \lambda^2(V^*),$$

and

$$\lambda^4 \otimes \lambda^2 = \lambda^4(V) \otimes \lambda^2(V) \xrightarrow{\varphi} \lambda^6(V) = \lambda^2(V^*),$$

where φ is multiplication in $\Lambda(V)$;

$$\lambda^4 \otimes \lambda^{2*} = \lambda^4(V^*) \otimes \lambda^2(V^*) \xrightarrow{\varphi} \lambda^6(V^*) = \lambda^2(V).$$

Theorem 12.1. *These maps embed* A *as a Lie algebra of maps from* $W \to W$ *and this action coincides up to isomorphism with the action of* $L(E_7)$ *on its 56-dimensional representation.*

Proof. We have maps $L \to Hom(\lambda^2, \lambda^2)$, $L \to Hom(\lambda^{2*}, \lambda^{2*})$, $\lambda^4 \to Hom(\lambda^2, \lambda^{2*})$, $\lambda^4 \to Hom(\lambda^{2*}, \lambda^2)$ which are all non-zero irreducible representations of $L(SL(V))$, so mono (by Schur's Lemma). Thus A embeds as stated. To check the action, choose a positive-definite Hermitian form on V compatible with $\lambda^n(V) \cong \mathbf{C}$. This gives a compact subgroup $SU(V) \subset SL(V)$ which we identify with the explicit $SU(8) \to E_7$. Then there is an isomorphism $\lambda^2 + \lambda^{2*} \to W'$ (the space of the 56-dimensional representation, Corollary 8.2 and Theorem 11.1) preserving the action of $SU(8)$. More explicitly, the action of $L = L(SU(8))$ on $\lambda^2 + \lambda^{2*}$ corresponds to the action of $L(SU(8)) \hookrightarrow L(E_7)$ on W'. Similarly, there exists an isomorphism $L + \lambda^4 \to L(E_7)$ which is the identity on L and on λ^4 preserves the action of $SU(8)$. The action of $L(E_7)$ on W' must correspond under this isomorphism to a sum of components; we want to know the components involving λ^4. The action of λ^4 must be

$$\lambda^4 \otimes \lambda^2 \xrightarrow{0} \lambda^2, \ \lambda^4 \otimes \lambda^2 \xrightarrow{c\varphi} \lambda^{2*}, \ \lambda^4 \otimes \lambda^{2*} \xrightarrow{c'\varphi} \lambda^2, \ \lambda^4 \otimes \lambda^{2*} \xrightarrow{0} \lambda^{2*},$$

where $c, c' \in \mathbf{C}$, as these are irreducible representations of $SU(8)$. We can change $\lambda^4 \to L(E_7)$, $\lambda^2 \to W'$, $\lambda^{2*} \to W'$ by multiplying by scalars γ, α, β, respectively. This replaces c by $(\alpha\gamma/\beta)c$, c' by $(\beta\gamma/\alpha)c'$ and by choosing α, β, γ we can arrange that $c = c' = 1$, so the action agrees up to isomorphism with the action of $L(E_7)$ on W'. This finishes the proof. □

However, to show that our present construction can be used to construct E_7 *ab initio*, we give a direct proof that the vector space of maps $Hom(W, W)$ is closed under taking commutators.

We begin by giving some brackets in A and checking that they agree with those in $Hom(W,W)$. Some of this will work if $\dim V = 4m$, so set $\dim V = 4m$, $A = L + \lambda^{2m}$ and on A put a symmetric form $(\ ,\ )$ with the following components. If $X, Y \in L$ are matrices, $(X,Y) = Tr(XY)$. Make L, λ^{2m} orthogonal and on λ^{2m}, choose $\lambda^{2m} \otimes \lambda^{2m} \xrightarrow{\varphi} \lambda^{4m} \cong \mathbf{C}$. Next, we construct a skew-symmetric bilinear product $[\ ,\ ] : (L + \lambda^{2m}) \otimes (L + \lambda^{2m}) \to L + \lambda^{2m}$. Well, the components $L \otimes L \to L$ and $L \otimes \lambda^{2m} \to \lambda^{2m}$ are respectively the usual Lie product and the action of L on λ^{2m}. We define $[\ ,\ ] : \lambda^{2m} \otimes \lambda^{2m} \to L$ by the equation $(a, [x,y]) = ([a,x], y)$, $a \in L$, $x, y \in \lambda^{2m}$. This characterises $[\ ,\ ]$ as $(\ ,\ )$ is non-singular on L.

We check that $[\ ,\ ]$ is skew-symmetric. For all $a \in L$, $x, y \in \lambda^{2m}$ we have

$$(a, [x,y]) + (a, [y,x]) = ([a,x], y) + ([a,y], x) = ([a,x], y) + (x, [a,y]) = 0,$$

because the bilinear form $(\ ,\ )$ on λ^{2m} is invariant under L. All this is strictly comparable with what we did in constructing E_8. We proceed to give an explicit formula for the bracket $\lambda^{2m} \otimes \lambda^{2m} \to L$, cf. Lemma 6.2.

Lemma 12.2. *Let $\{e_1, \ldots, e_{4m}\}$ be a basis of determinant 1 in V. Then,*

(i) $[e_{r_1} e_{r_2} \ldots e_{r_{2m}},\ e_{s_1} e_{s_2} \ldots e_{s_{2m}}] = 0$ *if two or more r's are s's,*

(ii) $[e_1 e_2 \ldots e_{2m},\ e_{2m} e_{2m+1} \ldots e_{4m-1}] = e_{2m} e^*_{4m}$,

(iii) $[e_1 e_2 \ldots e_{2m},\ e_{2m+1} \ldots e_{4m}] = \frac{1}{2}\left(\sum_{r=1}^{2m} e_r e^*_r - \sum_{r=2m+1}^{4m} e_r e^*_r\right)$.

Proof. (i) Let $a \in L(SL(V))$, $x = e_{r_1} \ldots e_{r_{2m}}$, $y = e_{s_1} \ldots e_{s_{2m}}$ and suppose two of the r's are equal to s's, say r', r''. Then $[a,x]$ is a sum of terms each of which is a multiple either of $e_{r'}$ or $e_{r''}$, so all of them annihilate y and $([a,x], y) = 0$. This holds for all a, so $[x,y] = 0$.

(ii) Let $a \in L(SL(V))$, $x = e_1 \ldots e_{2m}$, $y = e_{2m} \ldots e_{4m-1}$. Then $[a,x]$ contains terms each of which is divisible by e_{2m} and so annihilates e_{2m}, except for the term $e_1 \ldots e_{2m-1}(ae_{2m})$. Thus

$$([a,x], y) = e_1 \ldots e_{2m-1}(ae_{2m}) e_{2m} \ldots e_{4m-1} = e^*_{4m}(ae_{2m}).$$

This has the form $Tr(ab)$ where $b = e_{2m} e^*_{4m}$.

(iii) Let $a \in L(SL(V))$, $x = e_1 \ldots e_{2m}$, $y = e_{2m+1} \ldots e_{4m}$. Then

$$\begin{aligned} [a,x] &= \sum_{1 \le r \le 2m} e_1 \ldots (ae_r) \ldots e_{2m}, \qquad \text{and} \\ ([a,x], y) &= \sum_{1 \le r \le 2m} e_1 \ldots (ae_r) \ldots e_{2m} e_{2m+1} \ldots e_{4m} \\ &= \sum_{1 \le r \le 2m} e^*_r(ae_r). \end{aligned}$$

This has the form $Tr(ab)$ for $b = \lambda I + \sum_{1\le r\le 2m} e_r e_r^*$. Here λ is a scalar which we have to choose so as to get a matrix b of trace 0. Taking $\lambda = -\frac{1}{2}$ will do. □

For a general value of $\dim V$, there is no reason why this bracket should satisfy the Jacobi identity. However, we are studying the case $\dim V = 8$. We must check, without relying on the previous chapter, that the subspace $L + \lambda^4 \subset Hom(\lambda^2 + \lambda^{2*},\ \lambda^2 + \lambda^{2*})$ is closed under the formation of commutators $ab - ba$ and these commutators agree with the formula we have given in $L + \lambda^4$. That is, we wish to check that $x(yz) - y(xz) = [x, y]z$ for all $x, y \in L + \lambda^4$, $z \in \lambda^2 + \lambda^{2*}$. When $x, y \in L$ the result is true: the action of L on $\lambda^2 + \lambda^{2*}$ is a Lie action. Next, consider the case $x \in L$, $y \in \lambda^4$. Here the result is also true because the map $\lambda^4 \otimes (\lambda^2 + \lambda^{2*}) \to \lambda^2 + \lambda^{2*}$ is invariant under L. Explicitly, this means $x(yz) = [x, y]z + y(xz)$ for $x \in L$, $y \in \lambda^4$, $z \in \lambda^2 + \lambda^{2*}$. The case $y \in L$, $x \in \lambda^4$ follows because the result is skew-symmetric in x, y. So it remains to consider the case $x, y \in \lambda^4$. Because of the symmetry between V and V^*, we may take $z \in \lambda^2$: the other case is the same. We now have a map

$$f : \lambda^4 \otimes \lambda^4 \otimes \lambda^2 \to \lambda^2,$$

which is invariant under $L(SL(V))$, namely

$$x \otimes y \otimes z \mapsto x(yz) - y(xz) - [x, y]z.$$

We wish to prove this is zero. By the standard operating procedure, we can assume $x = e_a e_b e_c e_d$, $y = e_f e_g e_h e_i$, $z = e_j e_k$, where $a, b, \ldots,\ k \in \{1, \ldots, 8\}$ and $a, \ldots, d$ are distinct, $f, \ldots, i$ are distinct, and j, k are distinct. The matrix $diag(\zeta_1, \ldots, \zeta_8)$ acts on $x \otimes y \otimes z$ by multiplication with $\zeta_a \zeta_b \ldots \zeta_k$. Therefore, it acts on $f(x \otimes y \otimes z)$ in the same way. The result is 0 unless six of $\{1, \ldots, 8\}$ occur once among $a, \ldots, k$ and two, say, l, m occur twice; then the outcome will be a multiple of $e_l e_m$. The action of $N(T)$ allows us to permute the basis vectors in any way provided we alter the sign of one basis vector, if necessary. We then have the following cases to check.

Case (a)	$x = e_1e_2e_3e_4$,	$y = e_5e_6e_7e_8$,	$z = e_3e_4$, e_4e_8 or e_7e_8,
Case (b)	$x = e_1e_2e_3e_4$,	$y = e_4e_5e_6e_7$,	$z = e_3e_8$ or e_7e_8,
Case (c)	$x = e_1e_2e_3e_4$,	$y = e_3e_4e_5e_6$,	$z = e_7e_8$.

This completes the check that the subspace $L + \lambda^4$ is closed under commutators and the commutators are given by Lemma 12.2. But commutators always satisfy the Jacobi identity, so $A = L + \lambda^4$ is a Lie algebra, and we are finished with the discussion of Theorem 12.1. □

Addendum. *The inner product on A is invariant and the Killing form is* $36(\ ,\ )$.

Proof. The inner product is invariant under L by construction. Secondly, take $x \in L$, $y, z \in \lambda^4$. Then $(x, [y, z]) = ([x, y], z)$ by the definition of the Lie bracket. This equation gives $([y, x], z) + (x, [y, z]) = 0$, which shows that the form is invariant under $y \in \lambda^4$. For the Killing form, we argue, as we did for $L(E_8)$, that both the Killing form $(\ ,\ )_K$ and our chosen inner product $(\ ,\ )$ are invariant under $A = L + \lambda^4$ and A is irreducible, so we must have $(\ ,\ )_K = c(\ ,\ )$ for some scalar c. But on $L(SU(8))$ the trace of $z \mapsto [x, [y, z]]$ is $16Tr(xy)$ and the contribution of λ^4 is given as follows: for $x = e_1e_2e_ie_j,\ e_1e_ie_je_k,\ e_ie_je_ke_l$, $[e_1e_2^*, [x, e_2e_1^*]]$ is respectively $0,\ e_1e_ie_je_k,\ 0$ and so $Tr_{\lambda^4}(e_1e_2^* \cdot e_2e_1^*) = 20$. □

We go on to define further structure maps on W. Firstly, W is symplectic, so it has to carry a skew-symmetric bilinear form. Secondly, let us point out a principle that covers several known cases. (Compare [Al].) Suppose W is a faithful representation of a Lie group G, e.g. $Aut(A)$ on A for some Jordan algebra A. Then we have an embedding $i : L(G) \to Hom(W, W)$, for example, $Der(A) \to Hom(A, A)$. We also have $Hom(W, W) \cong W \otimes W^*$. If W is semi-simple, we expect a G-map from $W \otimes W^*$ back to $L(G)$ so that the composite is 1:

$$L(G) \xrightarrow{i} Hom(W, W) \cong W^* \otimes W \to L(G).$$

$L(G)$ will usually have an invariant product and $W \otimes W^*$ certainly does, so we can construct a map back by transposing i.

Apply this procedure to $W = \lambda^2 + \lambda^{2*}$. First put a skew-symmetric bilinear form on W with the following components

$$\lambda^2 \otimes \lambda^2 \xrightarrow{0} \mathbf{C},\ \lambda^2 \otimes \lambda^{2*} \xrightarrow{(\,,\,)} \mathbf{C},\ \lambda^{2*} \otimes \lambda^2 \xrightarrow{-(\,,\,)} \mathbf{C},\ \lambda^{2*} \otimes \lambda^{2*} \xrightarrow{0} \mathbf{C}.$$

Write this form as $\langle\ ,\ \rangle$. Then on $W \otimes W$ we have $(w_1 \otimes w_2, w_3 \otimes w_4) = \langle w_1, w_3 \rangle \langle w_2, w_4 \rangle$. This is symmetric and does not change sign if you interchange V and V^*.

Lemma 12.3. *There is a unique* $i : A \to W \otimes W$ *such that the composite* $A \otimes W \xrightarrow{i \otimes 1} W \otimes W \otimes W \xrightarrow{1 \otimes \langle\,,\,\rangle} W$ *is* $a \otimes w \longmapsto [a, w]$.

Proof. The action of A on W is a map $A \otimes W \to W$ which gives an injection $A \to Hom(W, W) = W \otimes W^* = W \otimes W$. The proof now follows from the fact that $\langle\ ,\ \rangle$ is non-singular. □

We now construct a symmetric map $W \otimes W \xrightarrow{\circ} A$ with the following components:

$$\lambda^2 \otimes \lambda^2 \xrightarrow{-\varphi} \lambda^4, \quad \lambda^{2*} \otimes \lambda^{2*} \xrightarrow{\varphi} \lambda^{4*} = \lambda^4$$

and

$$\lambda^2 \otimes \lambda^{2*} \longrightarrow V \otimes V^* \xrightarrow{\pi} L,$$

where the first map is

$$u_1u_2 \otimes v_1^*v_2^* \longmapsto (v_1^*, u_1)u_2 \otimes v_2^* - (v_1^*, u_2)u_1 \otimes v_2^* - (v_2^*, u_1)u_2 \otimes v_1^* + (v_2^*, u_2)u_1 \otimes v_1^*,$$

and $\pi(x) = x - \frac{1}{8}Tr(x)I$, after using $V \otimes V^* \cong Hom(V, V)$.

Theorem 12.4. *The form $\langle\ ,\ \rangle$, the product $\circ$ and the map i are invariant under A. The map i maps into the symmetric tensors and the diagram*

$$\begin{array}{ccccc} & & W \otimes W & & \\ & {\scriptstyle i}\nearrow & & \searrow{\scriptstyle \circ} & \\ A & & \xrightarrow{\quad 12 \quad} & & A \end{array}$$

commutes. For all $a \in A$, $y, z \in W$,

$$(a,\ y \circ z)_A = (ia,\ y \otimes z)_{W \otimes W} = \langle\, [a, y],\ z\rangle_W.$$

Proof. All our constructions are invariant under L. To prove that the form is invariant under λ^4, we wish to check that $\langle\, [x, y], z\rangle + \langle y, [x, z]\,\rangle = 0$ for all $x \in \lambda^4$, $y, z \in W = \lambda^2 + \lambda^{2*}$. If y, z are in opposite summands, then both products are 0, so we can assume that y, z are in the same summand. Suppose $y, z \in \lambda^2$. Then the definitions are $\langle\, [x, y],\ z\rangle = -(xy)z \in \lambda^8 = \mathbf{C}$ and $\langle y, [x, z]\,\rangle = y(xz) \in \lambda^8 = \mathbf{C}$. These add up to 0 since $xy = yx$. If $y, z \in \lambda^{2*}$, the proof is similar: interchange V, V^*. It is clear that the map i is invariant under A since all its ingredients are. We now check the relation between $(ix, iy)_{W \otimes W}$ and $(x, y)_A$. Since A is an irreducible representation of A, we must have $(ix, iy)_{W \otimes W} = c(x, y)_A$ for some scalar c and to determine c, we check on a pair of elements. The operation $[e_1e_2^*,\ ]$ carries $e_2e_r \in \lambda^2$ to e_1e_r for $3 \le r \le 8$ and e_re_s to 0 otherwise; $e_1^*e_r^*$ to $-e_2^*e_r^*$ for $3 \le r \le 8$ and $e_r^*e_s^*$ to 0 otherwise. So we have

$$i(e_1e_2^*) = \sum_{r=3}^{8} e_1e_r \otimes e_2^*e_r^* - \sum_{r=3}^{8} e_2^*e_r^* \otimes e_1e_r.$$

Similarly

$$i(e_2e_1^*) = \sum_{r=3}^{8} e_2e_r \otimes e_1^*e_r^* - \sum_{r=3}^{8} e_1^*e_r^* \otimes e_2e_r\,,$$

so $(i(e_2e_1^*), i(e_1e_2^*)) = 12 = c$.

To check that $(x,\ y \circ z)_A = \langle\, [x, y],\ z\rangle$, consider first the case $x \in L$, $y \in \lambda^2$, $z \in \lambda^{2*}$. Then, the second map in the definition of $y \circ z$ does not affect the pairing with x and for $y = u_1u_2$, $z = v_1^*v_2^*$ we have

$$\begin{aligned} &(x, y \circ z) \\ &\quad = (v_1^*, u_1)(v_2^*, xu_2) - (v_1^*, u_2)(v_2^*, xu_1) - (v_2^*, u_1)(v_1^*, xu_2) + (v_2^*, u_2)(v_1^*, xu_1) \\ &\quad = (v_1^*v_2^*, u_1(xu_2) + (xu_1)u_2) = (v_1^*v_2^*, x(u_1u_2)) = \langle\, [x, y], z\rangle\,. \end{aligned}$$

Consider next the case $x \in L$, $y \in \lambda^{2*}$, $z \in \lambda^2$. Here

$$\begin{aligned}(x, y \circ z) = (x, z \circ y) &= \langle [x,z], y\rangle, && \text{by the above}\\ &= -\langle z, [x,y]\,\rangle, && \text{by invariance under } L\\ &= \langle\, [x,y], z\rangle && \text{as } \langle\ ,\ \rangle \text{ is skew symmetric.}\end{aligned}$$

Note that this working established that $\langle\, [x,y],\, z\rangle$ is symmetric in y and z.

Consider thirdly the case $x \in \lambda^4$, $y, z \in \lambda^2$. The definitions give $(x,\, y \circ z) = -x(yz) \in \lambda^8 = \mathbf{C}$, $\langle\, [x,y],\, z\rangle = -(xy)z \in \lambda^8 = \mathbf{C}$. This proves the relation in all cases.

Next, we notice that the definition of i is (for $x \in A$) $ix = \sum_\alpha x'_\alpha \otimes x''_\alpha \in W \otimes W$ where $[x,y] = \sum_\alpha x'_\alpha \langle x''_\alpha, y\rangle$ for all $y \in W$. This says that $\langle\, [x,y], z\rangle = \sum_\alpha \langle x'_\alpha, z\rangle\langle x''_\alpha, y\rangle = (ix, z \otimes y)$. We have already seen that the left hand side is symmetric in y, z so this gives

$$(x,\, y \circ z) = \langle\, [x,y],\, z\rangle = (ix,\, y \otimes z)_{W\otimes W}\,, \tag{$*$}$$

which shows that the map $\circ$ is invariant under A. It also shows that ix lies in the space of symmetric tensors because $(ix,\, y \otimes z) = (ix,\, z \otimes y)$.

Finally we rewrite $(*)$ as

$$(x,\ \circ\, t) = (ix, t)_{W\otimes W}, \quad (t \in W \otimes W),$$

so on substitution, we have $(x,\ \circ\, ix')_A = (ix, ix')_{W\otimes W} = 12(x,x')_A$. Since this is true for all x we get $\circ\, ix' = 12x'$. This finishes the proof of Theorem 12.4. □

Corollary 12.5. *The action of E_7 on W is faithful.*

Proof. We have $L(E_7)$ as a summand in $W \otimes W$. Since all the above maps preserve the action of $L(E_7)$ and E_7 is connected, they also preserve the action of E_7. If $g \in E_7$ fixes W, it fixes $W \otimes W$, so fixes $L(E_7)$ and $L(E_7 \times SU(2)) \subset L(E_8)$. But it also fixes the complementary summand $L(E_8)/L(E_7 \times SU(2)) \cong W \otimes \lambda'$, so g fixes $L(E_8)$. Thus $g = 1$ in E_8. □

E_7 as a group of maps of W.

Cartan showed that W carries an invariant quartic function — equivalently an invariant symmetric map $W \otimes W \otimes W \otimes W \to \mathbf{C}$. (This equivalence is analogous to the equivalence of quadratic forms and symmetric bilinear forms: b gives q by $q(x) = b(x,x)$ and q gives b by $b(x,y) =$ coefficient of $2\lambda\mu$ in $q(\lambda x + \mu y)$. Similarly for n-linear functions f — we get $F(x) = f(x, \ldots, x)$ an n-tic function and if $F(x)$ is n-tic we recover f by $f(x_1, \ldots, x_n) =$ coefficient of $n!\,\lambda_1 \ldots \lambda_n$ in $F(\lambda_1 x_1 + \cdots + \lambda_n x_n)$. This process is called *polarisation.* In general it is better to work with the symmetric multilinear f.)

We define an invariant symmetric quadrilinear function on W by

$$f(w\otimes x\otimes y\otimes z)=\tfrac{1}{3}\Big((w\circ x,\,y\circ z)+(w\circ y,\,x\circ z)+(w\circ z,\,x\circ y)\Big).$$

The next theorem says that we can recover $(w\circ x,\,y\circ z)$ from f.

Theorem 12.6. $(w\circ x,\,y\circ z)=f(w\otimes x\otimes y\otimes z)-\frac{1}{2}\langle w,y\rangle\langle x,z\rangle-\frac{1}{2}\langle w,z\rangle\langle x,y\rangle$.

The proof requires lemmas: the first helps us work out $(w\circ x,\,y\circ z)$.

Lemma 12.7. If $a,c\in\lambda^2$, $b^*,d^*\in\lambda^{2*}$ then

$$(a\circ b^*,\,c\circ d^*)=\tfrac{1}{2}(b^*,a)(d^*,c)+(b^*,c)(d^*,a)-(b^*d^*,ac).$$

Proof. Use the standard operating procedure. □

We can now evaluate f explicitly.

Lemma 12.8. *If $a,b,c,d\in\lambda^2$, $a^*,b^*,c^*,d^*\in\lambda^{2*}$ then*

$$\begin{aligned}
f(a\otimes b\otimes c\otimes d)&=abcd\in\lambda^8(V)\cong\mathbf{C},\\
f(a\otimes b\otimes c\otimes d^*)&=0,\\
f(a\otimes b\otimes c^*\otimes d^*)&=\tfrac{1}{2}(c^*,a)(d^*,b)+\tfrac{1}{2}(c^*,b)(d^*,a)-(c^*d^*,ab),\\
f(a\otimes b^*\otimes c^*\otimes d^*)&=0,\\
f(a^*\otimes b^*\otimes c^*\otimes d^*)&=a^*b^*c^*d^*\in\lambda^8(V^*)\cong\mathbf{C}.
\end{aligned}$$

□

Proof of Theorem 12.6. This follows from Lemma 12.8. □

Corollary 12.9. *Let G be the identity component of the group of linear maps $g:W\to W$ which preserve $\langle\ ,\ \rangle$ and Cartan's invariant quartic function. Then $G\cong E_7$ (and G also preserves the associated quadrilinear function).*

Proof. G is a Lie group because it is a closed subgroup of a linear group. We have a homomorphism $E_7\to G$ given by the action of E_7 on W and it is mono by Corollary 12.5. G acts on $Hom(W,W)$ by $g(\theta)=g\theta g^{-1}$.

Claim. *This action preserves the subspace $A\subset Hom(W,W)$.*

Proof. Invariance gives $(gw\circ gx,\,gy\circ gz)=(w\circ x,\,y\circ z)$ from Theorem 12.4. Thus $\langle\,[gw\circ gx,\,gy],\,gz\rangle=\langle\,[w\circ x,\,y],\,z\rangle=\langle g[w\circ x,\,y],\,gz\rangle$ for all gz and since $\langle\ ,\ \rangle$ is non-singular, we have $g[w\circ x,y]=[gw\circ gx,gy]$, so regarding these as linear maps defined on gy we get $g(w\circ x)g^{-1}=gw\circ gx\in Hom(W,W)$. But elements of the form $w\circ x$ generate A by Theorem 12.4. Hence $gAg^{-1}=A$. □

Clearly $\theta\mapsto g\theta g^{-1}$ preserves commutators, hence there is a map $G\to Aut(A)$ with kernel $=\{\pm 1\}$. (Proof. $g:W\to W$ is in the kernel if and only if g

commutes with all $\theta \in A$ so $g\theta g^{-1} = \theta \Rightarrow g = \lambda$ by Schur's Lemma because A is irreducible, and since g preserves $\langle\ ,\ \rangle$ we must have $\lambda = \pm 1$. Clearly the two maps $\pm 1 : W \to W$ do preserve all structures and yield the identity element of $Aut(A)$.)

Now consider $E_7 \to G \to$ (the identity component of $Aut(A)$). As the Killing form on A is non-singular, by the Addendum to Theorem 12.1, this composite is epi, and $1, i \in SU(8)/\{\pm 1\}$ form $\ker(G \to$ the identity component of $Aut(A))$. We have as a consequence that $E_7 \to G$ is epi (and mono by Corollary 12.5), so $E_7 \cong G$. □

Real forms of E_7.

First we deal with the compact form which is what we obtain as a subgroup of E_8. The invariant inner product we had on the real vector space $L(E_8)$ was positive definite; it has to restrict to an invariant inner product on the real vector space $L(E_7)$ and this must be a scalar multiple of the one we have already. The one we have already was chosen negative definite on $L(SU(8))$, so the multiple has to be negative. The complementary summand $L(E_7)/L(SU(8))$ gives a real vector space U of dimension 70 whose complexification is the space λ^4 considered above; on this space U the inner product of $L(E_8)$ is real and positive definite, therefore the inner product of $L(E_7)$ is real and negative definite. The Killing form has signature -133.

For the second form of E_7, we work through the construction of the last chapter, taking V to be a real vector space of dimension 8, L to be $L(SL(V))$ and interpreting λ^4 as a real vector space. We get a real Lie algebra $L + \lambda^4$. The Killing form is still 36 times the inner product, and it has signature 7 on L, 0 on λ^4, giving a total of 7. Under this construction we see that $\lambda^2 + \lambda^{2*}$ is a representation of this version of E_7.

A third possibility is to go back to the subgroup $Spin(12) \times Sp(1)/\mathbf{Z}_2 \subset E_7$ and consider the element $-1 \in Spin(12)$. It acts as 1 on $L(Spin(12) \times Sp(1))$ but as -1 on the complex representation $\Delta^+ \otimes \lambda^1$. Let its -1 eigenspace on the real space $L(E_7)$ be U'. Then $L(E_7) = L(Spin(12) \times Sp(1)) + U'$ and $L(Spin(12) \times Sp(1)) + iU'$ is a new real Lie algebra. Its Killing form is negative definite on the first summand, of dimension 69, positive definite on the second, of dimension 64, so the signature is -5.

A fourth possibility rests on the existence of E_6. We have a map $E_6 \times SU(3) \to E_8$. We can also construct a map $S^1 \times SU(2) \to SU(3)$ mapping the circle into $diag(z^2, z^{-1}, z^{-1})$ and $SU(2)$ onto the last two coordinates as usual. So we get $E_6 \times S^1 \times SU(2) \to E_8$. Since $E_6 \times S^1$ centralises $SU(2)$, we have a map $E_6 \times S^1 \to E_7$. We wish to know the restriction of the representation $L(E_8)$. We have

$$L(SU(3)) \mapsto L(S^1 \times SU(2)) + (\eta^3 + \eta^{-3})\lambda_2^1,$$

where η is the identity representation of S^1; $\lambda_3^1 \mapsto \eta^2 + \eta^{-1}\lambda_2^1$, $\lambda_3^2 \mapsto \eta^{-2} + \eta\lambda_2^1$.

Thus letting M be the 27-dimensional representation of E_6, we get

$$L(E_8) \mapsto L(E_6 \times S^1 \times SU(2)) + (\eta^3 + \eta^{-3})\lambda_2^1 + M \otimes (\eta^2 + \eta^{-1}\lambda_2^1) + \overline{M} \otimes (\eta^{-2} + \eta\lambda_2^1).$$

Taking the part fixed under $SU(2)$, we see that

$$L(E_7) \mapsto L(E_6 \times S^1) + M \otimes \eta^2 + \overline{M} \otimes \eta^{-2}.$$

The element $i \in S^1$ acts as 1 on $L(E_6 \times S^1)$ and as -1 on the complementary summand. Let the -1 eigenspace of i acting on the real space $L(E_7)$ be U''. Then $L(E_7) = L(E_6 \times S^1) + U''$ and $L(E_6 \times S^1) + iU''$ is a new Lie algebra over $\mathbf{R}$; its complexification is still $\mathbf{C} \otimes L(E_7)$ and the signature of the Killing form is $-79 + 54 = -25$.

Chapter 13

Direct treatment of E_6

The object of this chapter is to do for E_6 what Chapter 12 did for E_7. We use three vector spaces V_1, V_2, V_3 of dimension 3 over $\mathbf{C}$ and their duals V_1^*, V_2^*, V_3^*. Choose an isomorphism $\lambda^3(V_i) \cong \mathbf{C}$, its dual $\lambda^3(V_i^*) \cong \mathbf{C}$ and use these to make the identifications $\lambda^2(V_i) \cong V_i^*$ and $\lambda^2(V_i^*) \cong V_i$. For $v, w \in V_i$, we have an anti-symmetric bilinear product $vw \in V_i^*$ and similarly for $v^*, w^* \in V_i^*$.

Lemma 13.1. *We have*

$$\begin{aligned} u^*(vw) &= (u^*, w)v - (u^*, v)w, \\ (vw)u^* &= (u^*, v)w - (u^*, w)v, \\ u(v^*w^*) &= (w^*, u)v^* - (v^*, u)w^*, \\ (v^*w^*)u &= (v^*, u)w^* - (w^*, u)v^*. \end{aligned}$$

Proof. For $(\mathbf{R}^3)^* = \mathbf{R}^3$ these are well-known properties of the vector product and they are proved using the standard operating procedure. □

Construction of E_6 and its 27-dimensional representation.

Everything we do will be invariant under $SL(V_i)$ which we identify with $SL(V_i^*)$. Put

$$\begin{aligned} L_i &= L(SL(V_i)) = L(SL(V_i^*)), \\ L &= L_1 + L_2 + L_3 = L(SL(3, \mathbf{C})^3) = L(SU(3)^3). \end{aligned}$$

As a candidate for $L(E_6)$, set

$$A = L + V_1 \otimes V_2 \otimes V_3 + V_1^* \otimes V_2^* \otimes V_3^* = L + V_1V_2V_3 + V_1^*V_2^*V_3^*,$$

dropping the $\otimes$, and as candidates for the 27-dimensional representations,

$$W = V_1V_2^* + V_2V_3^* + V_3V_1^*,$$

$$W^* = V_1V_3^* + V_2V_1^* + V_3V_2^*.$$

(Compare Theorem 11.4.) We may also identify $V_iV_j^*$ with $Hom_{\mathbf{C}}(V_j, V_i)$ and hence W and W^* with linear maps of $V_1 + V_2 + V_3$. Thus W is the space of matrices $\begin{pmatrix} 0 & A & 0 \\ 0 & 0 & B \\ C & 0 & 0 \end{pmatrix}$ and W^* is the space of matrices $\begin{pmatrix} 0 & 0 & D \\ E & 0 & 0 \\ 0 & F & 0 \end{pmatrix}$ where the blocks are 3×3. The duality of W and W^* is displayed by the map $W \otimes W^* \to \mathbf{C}$ given by $w \otimes w^* \mapsto Tr(w^*w)$ where w, w^* are viewed as the matrices above, or by $u_iu_j^* \otimes v_jv_i^* \mapsto (u_j^*, v_j)(v_i^*, u_i)$. The action of A on W and W^* is given as follows: $L = L_1 + L_2 + L_3$ acts as $L(SL(V_i)) = L(SL(V_i^*))$ acts on V_i and V_i^*; $u_1u_2u_3 \in V_1V_2V_3$ acts on $v_iv_j^* \in V_iV_j^*$ to give $(v_j^*, u_j)u_k \otimes u_iv_i \in V_kV_i^*$ $(k \neq i, j)$. Similarly $u_1^*u_2^*u_3^* \in V_1^*V_2^*V_3^*$ acts on $v_iv_j^* \in V_iV_j^*$ to give $(u_i^*, v_i)u_j^*v_j^* \otimes u_k^* \in V_jV_k^*$ $(k \neq i, j)$.

Lemma 13.2. *The pairing $W^* \otimes W \to \mathbf{C}$ is invariant under this action.*

Proof. Direct from the anti-symmetry of the triple bracket $\lambda^3(V_i) \cong \mathbf{C}$. □

Theorem 13.3. *These maps embed A as a Lie algebra of linear maps $W \to W$ (and $W^* \to W^*$). This action coincides, up to isomorphism, with the action of $L(E_6)$ on one of its 27-dimensional representations.*

Proof. Embedding: (As for E_7, see Theorem 12.1.) A is the sum of inequivalent irreducible representations L_1, L_2, L_3, $V_1V_2V_3$ and $V_1^*V_2^*V_3^*$ of $\prod_i SL(V_i)$ and the map into $Hom(W, W)$ is non-zero on each.

Next we show that the image of $A \to Hom(W, W)$ is closed under commutators.

It is clear that this is true if a or $b \in L$ because all our constructions are invariant under L. We see that the bracket in A of an element in L and an element in $V_1V_2V_3$ is given by the usual action of L on $V_1V_2V_3$. Suppose now $t_1t_2t_3, u_1u_2u_3 \in V_1V_2V_3$. We wish to define $[x, y]$ where $x = t_1t_2t_3, y = u_1u_2u_3$. If L is to act as a Lie algebra on W, we must have $[x, y]v = x(yv) - y(xv)$, for $v \in W$. The definitions give

$$[t_1t_2t_3,\ [u_1u_2u_3,\ v_iv_j^*]\,] = (v_j^*,\ u_j)[t_1t_2t_3,\ u_k \otimes u_iv_i] = (v_j^*,\ u_j)(t_iu_iv_i)t_j \otimes t_ku_k.$$

Interchanging the t's and u's gives

$$[u_1u_2u_3,\ [t_1t_2t_3,\ v_iv_j^*]\,] = (v_j^*,\ t_j)(u_it_iv_i)u_j \otimes u_kt_k = (v_j^*,\ t_j)(t_iu_iv_i)u_j \otimes t_ku_k.$$

The obvious element to try in $V_1^*V_2^*V_3^*$ is $(t_1u_1)(t_2u_2)(t_3u_3)$ and the action of this on $v_iv_j^*$ is:

$$\begin{aligned}[(t_1u_1)(t_2u_2)(t_3u_3),\ v_iv_j^*] &= (t_iu_iv_i)(t_ju_j)v_j^* \otimes t_ku_k \\ &= (t_iu_iv_i)(v_j^*,\ t_j)u_j \otimes t_ku_k - (t_iu_iv_i)(v_j^*,\ u_j)t_j \otimes t_ku_k,\end{aligned}$$

where we have used Lemma 13.1. We see that we should take

$$[t_1t_2t_3,\, u_1u_2u_3] = -(t_1u_1)(t_2u_2)(t_3u_3).$$

The case when t_i, u_i are replaced by t_i^*, u_i^* is exactly similar.

Finally take $t_1^*t_2^*t_3^* \in V_1^*V_2^*V_3^*$, $u_1u_2u_3 \in V_1V_2V_3$. From these elements we get an obvious element in L and its component in L_i is $(t_j^*,\, u_j)(t_k^*,\, u_k)\pi(u_i \otimes t_i^*)$, $j \neq k$, $j \neq i$, $k \neq i$, where $\pi : V_i \otimes V_i^* \to L_i$ is the projection defined by $\pi(x) = x - \frac{1}{3}(Tr\, x)I$.

The commutators in $Hom(W,W)$ satisfy the Jacobi identity, therefore so do the ones in A and we conclude that A is a Lie algebra.

To identify A with $L(E_6)$ and W with the 27-dimensional representation, we choose Hermitian forms on $V_1V_2V_3$, obtain groups $SU(V_i)$ and identify these with three copies of $SU(3)$ in $SU(3)^3 \to E_6$.

Now choose embeddings $V_1V_2V_3 \to L(E_6)$, $V_1V_2^*$, $V_2V_3^*$, $V_3V_1^* \to W'$ (the 27-dimensional representation) compatible with the action of $SU(3)^3$. We still have the freedom to multiply these embeddings by $\alpha, \beta, \gamma, \delta \in \mathbf{C}^*$ respectively. The action of $V_1V_2V_3$ (in $L(E_6)$) on these parts of W' must coincide with the above constructed action on $V_iV_j^*$ up to scalars (which are non-zero). We can vary these scalars by $\alpha\beta/\delta$, $\alpha\gamma/\beta$, $\alpha\delta/\gamma$ and so choose α to get the product of the three scalars to be 1. Choose β, γ, δ to get all of them 1. We thus identify W and W' so as to match up the action of L and $V_1V_2V_3$. Since A is generated by L and $V_1V_2V_3$ under brackets, it follows that $A \subset Hom(W,W)$ is contained in $Im(L(E_6)) \subset Hom(W',W') = Hom(W,W)$ and $Im(L(E_6))$ is closed under brackets. But $\dim A = 78$ and $\dim(Im(L(E_6))) \leq 78$. Hence $A = Im(L(E_6))$. □

Next we look for whatever other structure there is around. We have a symmetric bilinear form on A: on each L_i, it is to take matrices X, Y to $Tr(XY)$. For elements $u_1^*u_2^*u_3^* \in V_1^*V_2^*V_3^*$, $v_1v_2v_3 \in V_1V_2V_3$, we set

$$(u_1^*u_2^*u_3^*,\, v_1v_2v_3) = -\,(u_1^*,\, v_1)(u_2^*,\, v_2)(u_3^*,\, v_3).$$

The remaining summands are to be orthogonal.

Lemma 13.4. This symmetric bilinear form (,) on A is invariant and the Killing form is 24(,).

Proof. Straightforward. □

In the following lemma we use this inner product to transpose structures, compare Theorem 12.4.

Lemma 13.5. *There is a unique map* $W \otimes W^* \xrightarrow{\circ} A$ *such that* $(a, w \circ w^*) = ([a, w], w^*)$. *It is invariant under* A *and we have a commutative diagram:*

$$\begin{array}{ccc} & & Hom(W,W) \cong W \otimes W^* \\ & \nearrow & \downarrow \circ \\ A & \xrightarrow{\quad 6 \quad} & A \end{array}$$

Proof. Since the inner product in A is non-singular, it is clear that this map exists and is unique. It is invariant since all the ingredients of the structure are. For later use we give the map explicitly as follows:

on $V_1V_2^* \otimes V_2V_1^*$, we have $u_1u_2^* \circ v_2v_1^* = (u_2^*, v_2)\pi(u_1 \otimes v_1^*) - (v_1^*, u_1)\pi(v_2 \otimes u_2^*)$,
where $\pi(x) = x - \frac{1}{3}Tr(x)I$ and $u_i \otimes v_i^*$ is to be considered as a matrix,
on $V_1V_2^* \otimes V_3V_2^*$, we have $u_1u_2^* \circ v_3v_2^* = -\, u_1 \otimes u_2^*v_2^* \otimes v_3 \in V_1V_2V_3$, and
on $V_1V_2^* \otimes V_1V_3^*$, we have $u_1u_2^* \circ v_1v_3^* = -\, u_1v_1 \otimes u_2^* \otimes v_3^* \in V_1^*V_2^*V_3^*$. □

Corollary. *The action of* E_6 *on either of its 27-dimensional representations is faithful.*

Proof. (As for E_7, Corollary 12.5.) Suppose $g \in E_6 \subset E_8$ fixes W, the 27-dimensional representation. Then by duality it fixes W^* so also fixes $W \otimes W^*$. But $L(E_6)$ is a summand of $W \otimes W^*$, therefore g fixes $L(E_6)$ and $L(E_6 \times SU(3)) \subset L(E_8)$; it also fixes the complementary summand $L(E_8)/L(E_6 \times SU(3)) \cong W \otimes \lambda^1 + W^* \otimes \lambda^2$, so it fixes $L(E_8)$. Thus $g = 1$ in E_8. □

Next we define two bilinear maps $W \otimes W \xrightarrow{\times} W^*$, $W^* \otimes W^* \xrightarrow{\times} W$ as follows:

on $V_iV_j^* \otimes V_iV_j^*$, $u_iu_j^* \times v_iv_j^* = -(u_j^*v_j^*) \otimes (u_iv_i) \in V_jV_i^*$,

on $V_iV_j^* \otimes V_jV_k^*$ $(i, j, k$ distinct$)$, $u_iu_j^* \times v_jv_k^* = (u_j^*, v_j)u_i \otimes v_k^* \in V_iV_k^*$, and

on $V_jV_k^* \otimes V_iV_j^*$, $v_jv_k^* \times u_iu_j^* = (u_j^*, v_j)u_i \otimes v_k^* \in V_iV_k^*$.

Hence $\times$ is symmetric. We can also express this as follows.

If A is a 3×3 matrix, then $Adj\,A$ is a quartic function of A so we may polarise: $Adj(\mu A + \nu B) = \mu^2 Adj A + 2\mu\nu(A \times B) + \nu^2 Adj B$. The product on matrices is defined by

$$\begin{pmatrix} 0 & A & 0 \\ 0 & 0 & B \\ C & 0 & 0 \end{pmatrix} \times \begin{pmatrix} 0 & D & 0 \\ 0 & 0 & E \\ F & 0 & 0 \end{pmatrix}$$

$$= \begin{pmatrix} 0 & 0 & AE + DB - C \times F \\ BF + EC - A \times D & 0 & 0 \\ 0 & CD + FA - B \times E & 0 \end{pmatrix}$$

Lemma 13.6. *These products are invariant under $L(E_6)$, therefore also under E_6.* □

The following lemma says we can recover the inner product $W^* \otimes W \to \mathbf{C}$ from $\times$.

Lemma 13.7. *The trace of $W \xrightarrow{w\times} W^* \xrightarrow{w^*\times} W$ is $10(w^*, w)$ and similarly for $W^* \xrightarrow{w^*\times} W \xrightarrow{w\times} W^*$.* □

We proceed to introduce trilinear functions $f : W \otimes W \otimes W \to \mathbf{C}$ and $f : W^* \otimes W^* \otimes W^* \to \mathbf{C}$ by $f(x \otimes y \otimes z) = (x \times y,\, z)$.

Lemma 13.8. *The functions f are symmetric in all three variables.*

Proof. On $V_i V_j^* \otimes V_i V_j^* \otimes V_i V_j^*$, we have $f(u_i u_j^* \otimes v_i v_j^* \otimes w_i w_j^*) = -(u_i v_i w_i)(u_j^* v_j^* w_j^*)$, where both factors are anti-symmetric. Otherwise a summand on which f is non-zero is, up to rearrangement of factors, $V_1^* V_1 V_2^* V_2 V_3^* V_3$, and $f(v_1^* v_1 \otimes v_2^* v_2 \otimes v_3^* v_3) = (v_1^*, v_1)(v_2^*, v_2)(v_3^*, v_3)$. □

Since the trilinear functions f are symmetric, we lose no information by considering the associated cubic function $c(x) = f(x \otimes x \otimes x)$. We give it in matrix notation.

Lemma 13.9.

$$c\begin{pmatrix} 0 & A & 0 \\ 0 & 0 & B \\ C & 0 & 0 \end{pmatrix} = 6\,(Tr(ABC) - \det A - \det B - \det C),$$

$$c\begin{pmatrix} 0 & 0 & D \\ E & 0 & 0 \\ 0 & F & 0 \end{pmatrix} = 6\,(Tr(FED) - \det D - \det E - \det F). \quad □$$

Everything else can be reconstructed from the two cubic functions and the pairing, but as polarisation is inconvenient, it is better to retain formulae for the cross products. The next lemma is the key in recovering everything from the cross product.

Lemma 13.10. *For $x, z \in W$, $y^*, z^* \in W^*$ we have*

$$[x \circ y^*,\, z] = (y^*, z)x + \tfrac{1}{3}(y^*, x)z - (x \times z) \times y^*, \qquad \text{and}$$
$$[x \circ y^*,\, z^*] = -(z^*, x)y^* - \tfrac{1}{3}(y^*, x)z^* + x \times (y^* \times z^*).$$

Proof. It is only necessary to prove one formula as the other follows by transposing the first using the fact that (z, z^*) is invariant under the action of $x \circ y^*$. The verification of either formula is tedious. To check the coefficients on the right-hand side, note that $[x \circ y^*,\, z]$ is a linear function of z and so the trace of this linear function should be 0. On the right-hand side, the trace of $z \mapsto z$ is 27 and that of $z \mapsto (x \times z) \times y^*$ is $10(y^*,\, x)$, by Lemma 13.7. ∎

At this point, starting with W, W^* we have a possible alternative way to construct a Lie algebra of maps $W \to W$. Start with the products $\times : W \otimes W \to W^*$, $\times : W^* \otimes W^* \to W$, and recover the pairing $W^* \otimes W \to \mathbf{C}$ as before. So we know the right-hand sides of Lemma 13.10. Unfortunately, with this construction, to check that we have the required properties, we need some more properties of the cross product. From the cross product we can construct several maps from $W^4 \to W$ with various symmetries, but up to a scalar, we have only one map symmetric in all four variables.

Lemma 13.11. *For $w, x, y, z \in W$, we have a symmetric map $W^4 \to W$ given by $x \otimes y \otimes z \otimes w \mapsto$*

$$\begin{aligned} f(x \otimes y \otimes z)w + f(w \otimes y \otimes z)x + f(w \otimes x \otimes z)y + f(w \otimes x \otimes y)z \\ = (w \times x) \times (y \times z) + (w \times y) \times (x \times z) + (w \times z) \times (x \times y). \end{aligned}$$

Similarly for four variables in W^*.

Proof. $W \otimes W \xrightarrow{\times} W^*$ is invariant under the action of $[x \circ y^*,\ ]$, so

$$[x \circ y^*,\, w \times z] = [x \circ y^*,\, w] \times z + w \times [x \circ y^*,\, z],$$

or

$$[x \circ y^*,\, w] \times z + w \times [x \circ y^*,\, z] - [x \circ y^*,\, w \times z] = 0.$$

On substituting values from the formulae of Lemma 13.10, the left-hand side of this equation gives

$$\begin{aligned} &(y^*, w)\, x \times z + \tfrac{1}{3}(y^*, x)\, w \times z - ((x \times w) \times y^*) \times z \\ &\quad + (y^*, z)\, w \times x + \tfrac{1}{3}(y^*, x)\, w \times z - w \times ((x \times w) \times y^*) \\ &\quad + (w \times z,\, x)\, y^* + \tfrac{1}{3}(y^*, x)\, w \times z - (x \times (y^*(w \times z))) = 0 \in W^*. \end{aligned}$$

Take the inner product with $y \in W$ to get

$$\begin{aligned} &(y^*, z) f(w \otimes x \otimes y) + (y^*, x) f(w \otimes y \otimes z) \\ &\quad + (y^*, w) f(x \otimes y \otimes z) + (y, y^*) f(w \otimes x \otimes z) \\ &\quad - f(y \otimes w \otimes (x \times z) \times y^*) - f(y \otimes z \otimes (x \times w) \times y^*) \\ &\quad - f(x \otimes y \otimes (w \times z) \times y^*) = 0, \end{aligned}$$

which leads to the result when we remove y^*, since the above is true for all y^*. ∎

Lemma 13.12. *For all $x, y, w \in W$,*

$$w \circ (x \times y) + x \circ (y \times w) + y \circ (w \times x) = 0,$$

and for $w, x, y \in W^*$,

$$(w \times x) \circ y + (x \times y) \circ w + (y \times w) \circ x = 0.$$

Proof. Since $A \to Hom(W, W)$ is mono, it is sufficient to prove that the first expression gives 0 when it acts on any $z \in W$. When we substitute from Lemma 13.10 to find the value of $[s \circ t^*, z]$, we get an answer which vanishes by Lemma 13.11. Similarly for the second expression using the action on W^*. □

E_6 as a group of maps.

Finally, we wish to show that the symmetry group G of the structure we have found is exactly E_6. Our structure consists of two vector spaces W, W^* and two products $W \otimes W \xrightarrow{\times} W^*$ and $W^* \otimes W^* \xrightarrow{\times} W$. Then an element g will be a pair of linear maps $\gamma_1 : W \to W$, $\gamma_2 : W^* \to W^*$ such that for all $x, y \in W$, or $x, y \in W^*$, $g(x \times y) = gx \times gy$ where the action of g on W is by γ_1 and on W^* by γ_2.

Lemma 13.13. *The symmetry group G preserving these products is E_6.*

Proof. It is clear that an element g must preserve the pairing $W^* \otimes W \to \mathbf{C}$. In fact the trace of the composite

$$W \xrightarrow{w \times} W^* \xrightarrow{w^* \times} W$$

is equal to the trace of the composite

$$W \xrightarrow{(gw) \times} W^* \xrightarrow{(gw^*) \times} W,$$

that is $10\,(w^*, w) = 10\,(gw^*, gw)$, by Lemma 13.7.

Next, the subspace $A \subset Hom(W, W)$ is closed under $a \mapsto gag^{-1}$. To check this, from Lemma 13.10 we have

$$[x \circ y^*,\, z] = (y^*, z)x + \tfrac{1}{3}(y^*, x)z - (x \times z) \times y^*$$

and this gives

$$g[x \circ y^*,\, z] = (gy^*, gz)gx + \tfrac{1}{3}(gy^*,\, gx)gz - (gx \times gz) \times gy^* = [gx \circ gy^*,\, gz].$$

Thus

$$g[x \circ y^*,\, g^{-1}z] = [gx \circ gy^*,\, z],$$

for all $x, z \in W$, $y^* \in W^*$. So, for $a = x \circ y^*$ the map $a \mapsto gag^{-1}$ lies in A and as such a generate A, A is closed under $a \mapsto gag^{-1}$. The endomorphism $a \mapsto gag^{-1}$ of A preserves commutators, so we get a homomorphism $G \to Aut(A)$. Its kernel consists of the elements g such that the action on W commutes with the action of all $a \in A$. Since W is an irreducible representation of A, g must be multiplication by a complex scalar α and since g preserves the function $f(x \otimes y \otimes z)$, we must have $\alpha^3 = 1$ so $\alpha = 1, w, w^2$. Clearly these maps *do* preserve all structure and yield the identity element of $Aut(A)$. Consider the composite $E_6 \to G \to$ (the identity component of $Aut(A)$). The Killing form on $L(E_6)$ is non-singular by Lemma 13.4, so the map $E_6 \to$ (the identity component of $Aut(A)$) is epi. Moreover, there are enough elements in the kernel of this map to account for 1, w, w^2 in $\ker(G \to Aut(A))$. That is to say, we have an inclusion $Spin(10) \times U(1)/\mathbf{Z}_4 \to E_6$ and the elements of order 3 in $U(1)$ do what is required. Thus the map $E_6 \to G$ is epi. It is mono by the corollary of Lemma 13.5. □

Chapter 14

Direct treatment of F_4, I

In the previous two Chapters, we studied E_6 and E_7 by way of their lowest dimensional non-trivial representations. Here we do the same for F_4. We relate the Lie brackets in $L(E_8)$ to the brackets in $L(F_4), L(G_2)$ and structures on the 26-dimensional representation U and the 7-dimensional representation V of F_4 and G_2 respectively. We also construct an algebra structure on U. In Chapter 16 we will construct U, together with all accompanying structures, independently of this Chapter and show that F_4 is indeed the group of automorphisms of U preserving these structures.

F_4 contains $Spin(9)$ as a subgroup of maximal rank:

$$F_4 \supset Spin(9) \supset Spin(8) \supset T.$$

We use the usual notation of x_1, x_2, x_3, x_4 for weights. From Chapter 9, the roots of F_4 are

long roots :	$\pm x_i \pm x_j$,	$i < j$,	24 roots of $Spin(8)$,
short roots :	$\pm x_i$		8,
	$\frac{1}{2}(\pm x_1 \pm x_2 \pm x_3 \pm x_4)$		16.

Before going on it will be handy to introduce a particular set of subgroups of F_4.

Lemma 14.1. *The group F_4 has a subgroup of local type $SU(3) \times SU(3)$ so that the roots corresponding to the first factor $SU(3)$ are long roots of F_4 and the roots corresponding to the second factor are short roots.*
(Note that these $SU(3)$ are not in E_6.)

Proof. Consider $SU(3) \to Spin(6) \to Spin(9) \to F_4$. This $SU(3)$ has roots $\pm(x_i - x_j)$, $1 \leq i < j \leq 3$, and its $L(T)$ is determined by $x_1 + x_2 + x_3 = 0$,

$x_4 = 0$. Now consider its centraliser in F_4; its $L(T)$ is defined by $x_1 = x_2 = x_3$ and its roots are

$$\begin{array}{ll} \pm\frac{1}{2}(x_1+x_2+x_3+x_4) & \text{weights for } \Delta^+, \\ \pm\frac{1}{2}(x_1+x_2+x_3-x_4) & \qquad\qquad\quad \Delta^-, \\ \pm x_4 & \qquad\qquad\quad \lambda^1. \end{array}$$

This centraliser is of type $SU(3)$ which inherits its roots from the short roots of F_4. Thus we get a map $SU(3) \times SU(3) \to F_4$ as required. □

Theorem 14.2. *The Weyl group of F_4 has order $|W(F_4)| = 2^7 \cdot 3^2$; $W(Spin(8))$ is a normal subgroup of $W(F_4)$ with $W(F_4)/W(Spin(8)) \cong \Sigma_3$ and this extension is split. The quotient Σ_3 preserves $Spin(8)$ — indeed it is $Out(Spin(8))$. Thus all automorphisms of $Spin(8)$ become inner in F_4.*

Proof. Consider $x \in N(T) \subset F_4$. Then $y \mapsto xyx^{-1}$ gives an inner automorphism preserving T. Clearly it carries long roots to long and short roots to short, so preserves

$$L(Spin(8)) = L(T) + \langle\, e_\theta \mid \theta \text{ long} \,\rangle,$$

where e_θ is an eigenvector corresponding to θ. Therefore it preserves $Spin(8)$. In particular, it preserves $W(Spin(8))$ which shows that $W(Spin(8))$ is normal in $W(F_4)$.

To continue, we have a map

$$N(T) \to Aut(Spin(8))/Inn(Spin(8)) = Out(Spin(8)) = \Sigma_3.$$

Since clearly T maps to 1, we get a map

$$W(F_4) = N(T)/T \to Out(Spin(8)) = \Sigma_3.$$

The kernel is $W(Spin(8))$. In fact, it is clear that an element of $W(Spin(8))$ maps into $Inn(Spin(8))$. Conversely, suppose given an inner automorphism $y \mapsto xyx^{-1}$ of F_4 which restricts to an inner automorphism $y \mapsto zyz^{-1}$ of $Spin(8)$. Then its effect on T can be given by $y \mapsto zyz^{-1}$ in $Spin(8)$, i.e. we had an element of $W(Spin(8))$. Hence we get a monomorphism $W(F_4)/W(Spin(8)) \to \Sigma_3$. This map is split epi. For, consider the subgroups of type $SU(3) \times SU(3)$ of Lemma 14.1. We get $W(SU(3) \times SU(3)) \subset W(F_4)$ and in particular $W(F_4)$ contains a Σ_3 subgroup given by the Weyl group of the second factor, the one which inherits its roots from the short roots of F_4. This Σ_3 acts by permuting the root pairs

$$\pm\tfrac{1}{2}(x_1+x_2+x_3+x_4), \quad \pm\tfrac{1}{2}(x_1+x_2+x_3-x_4), \quad \pm x_4.$$

Therefore, this Σ_3 permutes the representations Δ^+, Δ^-, λ^1 of $Spin(8)$ in the same way. □

To continue, let U denote the 26-dimensional representation of F_4 and recall that

(i) U is real;

(ii) $U|\,Spin(9)$ is $1+\lambda^1+\Delta$;

(iii) as a representation of $F_4 \times G_2$ we have

$$L(E_8) = L(F_4) + L(G_2) + U \otimes V,$$

where V is the 7-dimensional representation of G_2 arising from $G_2 \subset Spin(7)$.

Thus $U|Spin(8)$ is $2+\lambda^1+\Delta^+ +\Delta^-$ and in particular, its weights are the short roots of F_4 together with zero (twice).

Lemma 14.3. *The representation* U *of* F_4 *is irreducible.*

Proof. The non-zero weights form a single orbit under $W(F_4)$ so the only possible decomposition for U is $U = 1 + U'$, where $U' \mapsto \lambda^1 + \Delta$ on $Spin(9)$. Restrict to the subgroups of local type $SU(3) \times SU(3)$ of Lemma 14.1. Then, by consideration of weights, we see that $U \mapsto 1 \otimes adj + \lambda^1 \otimes \lambda^1 + \lambda^2 \otimes \lambda^2$, where the parts are of dimension 8, 9, 9 respectively. The two zero weights are in $1 \otimes adj$ which, nevertheless, is an irreducible representation of $SU(3) \times SU(3)$. We see that U must be irreducible. □

Lemma 14.4. (i) *As a representation of* F_4, *we have* $L(E_6) = L(F_4)+U$, *so* U *can be constructed as a representation of* F_4 *on* $L(E_6)/L(F_4)$ *(giving another proof that* U *is real).*

(ii) *Both the 27-dimensional representations of* E_6 *restrict over* F_4 *to* $1+U$.

(iii) *On restriction from* E_6 *to* F_4, *24 of the 27 weights of each 27-dimensional representation restrict to the 24 short roots of* F_4. *The remaining 3 restrict to zero and these 3 add to zero before restriction.*

Proof. We have

$$\begin{aligned} L(E_8) &= L(F_4) + L(G_2) + U \otimes V && \text{over } F_4 \times G_2; \\ L(E_8) &= L(F_4) + L(SU(3)) + \lambda^1 + \lambda^2 + U \otimes (\lambda^1 + \lambda^2 + 1) && \text{over } F_4 \times SU(3); \\ L(E_8) &= L(E_6) + L(SU(3)) + W \otimes \lambda^1 + W^* \otimes \lambda^2 && \text{over } E_6 \times SU(3), \end{aligned}$$

where W and W^* are the 27-dimensional representations of E_6, see Chapter 13. Thus taking the part fixed under $SU(3)$, we get $L(E_6) = L(F_4) + U$, and taking the coefficients of λ^1 and λ^2 we get $W^*, W \mapsto 1 + U$, which proves (i), (ii). To finish the proof of (iii), we need only check that the three roots add to zero before restriction. With explicit calculation and our usual coordinates for F_4, we see that the three weights which restrict to 0 on F_4 are x_5+x_6, $-x_5+x_6$, $-2x_6$ for one representation and minus these for the other. (Note that x_6, x_7, x_8 all give the same function on $T \subset E_6$.) □

Structure maps on U.

We are now ready to see what sort of structure maps the representation U carries.

Theorem 14.5. (i) *The representation U carries a symmetric bilinear map $b : U \otimes U \to \mathbf{C}$ invariant under F_4 and non-singular.* (This follows from the reality of U.)

(ii) *The representation U carries a symmetric bilinear map $U \otimes U \xrightarrow{\times} U$ which is invariant under F_4 and epi.*

(iii) *Moreover, the function $U^3 \to \mathbf{C}, t(u', u'', u''') = b(u' \times u'', u''')$ is trilinear and symmetric in all three variables.*

Proof 1. This uses our knowledge of E_6.

Let W be the 27-dimensional representation of E_6. We know that there is a non-zero $f : W \otimes W \otimes W \to \mathbf{C}$, symmetric and invariant under E_6. By restricting to F_4 we get a function $f : (1+U) \otimes (1+U) \otimes (1+U) \to \mathbf{C}$, which is symmetric, invariant under F_4 and non-zero. All we need to know now is that $f : U \otimes U \otimes U \to \mathbf{C}$ is non-zero, for if it is and we define $\times$ by $b(u' \times u'', u''') = f(u' \otimes u'' \otimes u''')$, we get a product $\times : U \otimes U \to U$ invariant under F_4, non-zero and epi since U is irreducible. To show that $f \neq 0$, recall from Chapter 13 that $W = V_1V_2^* + V_2V_3^* + V_3V_1^*$. Let the weights of

V_1	be x_i,	$\sum x_i = 0$,
V_2	be y_i,	$\sum y_i = 0$,
V_3	be z_i,	$\sum z_i = 0$.

The weights of W are $x_i - y_j$, $y_j - z_k$, $z_k - x_i$.

The triples of weights adding to zero are

(i) $x_i - y_j$, $y_j - z_k$, $z_k - x_i$;

(ii) $x_i - y_j$, $x_k - y_\ell$, $x_m - y_n$, where $\{i,k,m\} = \{1,2,3\}$, $\{j,\ell,n\} = \{1,2,3\}$.

Whichever triple restricts to 0 on $T(F_4)$, we can find another such $\{\theta, \varphi, \psi\}$ disjoint from it. Then the corresponding eigenvectors e_θ, e_φ, e_ψ will lie in the summand U of $W = 1 + U$ and we have $f(e_\theta \otimes e_\varphi \otimes e_\psi) \neq 0$.

Proof 2. This requires some preparation and will use the fact that $F_4 = C_{E_8}(G_2)$, the centraliser of G_2 in E_8, see Chapter 8.

Lemma 14.6. *As representations of G_2, we have $\lambda^2(V) \cong L(G_2)+V$, where $G_2 \subset Spin(7)$ acts on $\mathbf{R}^7$ giving a 7-dimensional representation V.*

Proof. Restrict to $SU(3)$. Then

$$\begin{aligned} V \quad &\mapsto \lambda^1 + \lambda^2 + 1, \\ L(G_2) &\mapsto L(SU(3)) + \lambda^1 + \lambda^2, \\ \lambda^2(V) &\mapsto \lambda^2 + \lambda^1 + 0 + \lambda^1\lambda^2 + \lambda^1 + \lambda^2 \\ &\qquad = L(SU(3)) + 2\lambda^1 + 2\lambda^2 + 1 = L(G_2) + V. \end{aligned}$$

Thus $\lambda^2(V)$ and $L(G_2) + V$ are isomorphic over $SU(3)$, and as $SU(3)$ is of maximal rank in G_2, they are isomorphic over G_2. □

Note also that we have a product $V \otimes V \xrightarrow{\circ} L(G_2)$ given by $(a, v \circ w) = ([a,v], w)$. One checks that $\circ$ is anti-symmetric.

Corollary 14.7. *There is a skew symmetric product $V \otimes V \xrightarrow{\times} V$ invariant under G_2 and it is unique up to a scalar factor.* (This is obvious if we regard V as the space of pure imaginary Cayley numbers.)

Proof. The representation V is irreducible; in fact, from its weights, it could only split as $1 + V'$. We know that G_2 is transitive on $S^6 \subset V$. Also $L(G_2)$ is irreducible and disjoint from V so that we have a G_2-map $\lambda^2(V) \to V$, using Lemma 14.6. □

Lemma 14.8. *Any skew-symmetric map $\mu : U \otimes U \to U$ invariant under F_4 is zero.*

Proof. Restrict to $Spin(9)$. Then $U \mapsto 1 + \lambda_9^1 + \Delta$, so

$$\lambda^2(U) \mapsto \lambda^2 + \lambda^2(\Delta) + \lambda^1 + \Delta + \lambda^1 \otimes \Delta = \lambda^2 + (\lambda^2 + \lambda^3) + \lambda^1 + \Delta + \lambda^1 \otimes \Delta.$$

(See Theorem 4.6.) Thus U has a summand 1 and $\lambda^2(U)$ does not, so $\lambda^2(U)$ does not have a U summand. □

Now consider $L(E_8) = L(F_4) + L(G_2) + U \otimes V$. We put an inner product on $L(E_8)$, non-singular, symmetric, bilinear and invariant under E_8. The three summands are to be orthogonal. Moreover, since U and V are irreducible, there is, up to a scalar, only one way to put a bilinear form on $U \otimes V$: we must have, for $s,t \in U$, $v,w \in V$, $(s \otimes v, t \otimes w) = b(s,t)(v,w)_V$ where $(\ ,\ )_V$ is some inner product on V, invariant under G_2 and $b(\ ,\ )$ is a bilinear form on U, invariant under F_4. We know the Lie brackets on $L(F_4)$ and $L(G_2)$ and wish to determine other brackets. Well, $L(F_4)$ acts on $U \otimes V$ via its action on U and $L(G_2)$ acts on $U \otimes V$ via its action on V. There remains the Lie bracket $(U \otimes V) \otimes (U \otimes V) \to L(E_8)$. *A priori,* this has three components: one into $L(F_4)$, one into $L(G_2)$ and one into $U \otimes V$. For present purposes, we are interested in the third.

Lemma 14.9. *The component of the Lie bracket* $(U \otimes V)^2 \to U \otimes V$ *is*

$$[s \otimes v,\, t \otimes w] = (s \times t) \otimes (v \times w)$$

where $\times : V \otimes V \to V$ *is as in Corollary 14.7, and* $\times : U \otimes U \to U$ *is symmetric, invariant under* F_4 *and epi.*

Proof. The Lie bracket is a skew-symmetric map

$$(U \otimes V) \otimes (U \otimes V) \to U \otimes U \otimes V \otimes V \xrightarrow{\sum f_i \otimes g_i} U \otimes V,$$

for some $f_i : U \otimes U \to U$ invariant under F_4, symmetric by Lemma 14.8, and $g_i : V \otimes V \to V$ invariant under G_2 and skew-symmetric. By Corollary 14.7, each $g_i = \lambda_i g$ for some scalar λ_i, so $\sum f_i \otimes g_i = f \otimes g$ where $f = \sum \lambda_i f_i$. It remains to prove that $f \neq 0$. Recall that if θ, φ, $\theta + \varphi$ are roots of E_8 and e_θ, e_φ, $e_{\theta+\varphi}$ the corresponding eigenvectors in $L(E_8)$ over $\mathbf{C}$, then $[e_\theta, e_\varphi] = \lambda e_{\theta+\varphi}$ for some $\lambda \neq 0$. We want to choose θ, φ so that the restrictions to F_4 and G_2 are what we require. In fact, taking

$$\begin{aligned}\theta &= \tfrac{1}{2}(x_1 + x_2 + x_3 + x_4 + x_5 + x_6 - x_7 - x_8),\\ \varphi &= \tfrac{1}{2}(-x_1 - x_2 - x_3 + x_4 - x_5 - x_6 + x_7 - x_8),\end{aligned}$$

we have $\theta + \varphi = x_4 - x_8$, and e_θ, e_φ, $e_{\theta+\varphi}$ lie in $U \otimes V$ over $\mathbf{C}$. Hence $[\ ,\ ]$ is non-zero. This shows that $\times : U \otimes U \to U$ is non-zero. Since it is invariant under F_4 and U is irreducible, it must be epi. ☐

It remains to prove the third clause of Theorem 14.5, but we will prove a bit more. Let us use s, t, u for variables in U and v, w, x for variables in V.

Lemma 14.10. (i) $b(s \times t,\, u) = b(s,\, t \times u)$ *and* b *is symmetric in all three variables.*

(ii) $(v \times w,\, x)_V = (v,\, w \times x)_V$ *and is an alternating function of its three variables.*

Proof. The inner product in $L(E_8)$ is invariant, so

$$\begin{aligned}([s \otimes v,\, t \otimes w],\, u \otimes x) &= (s \otimes v,\, [t \otimes w,\, u \otimes x]), \qquad \text{that is}\\ ((s \times t) \otimes (v \times w),\, u \otimes x) &= (s \otimes v,\, (t \times u) \otimes (w \times x)).\end{aligned}$$

Hence

$$b(s \times t,\, u)(v \times w,\, x)_V = b(s,\, t \times u)(v,\, w \times x)_V. \qquad (*)$$

Thus the functions $s \otimes t \otimes u \mapsto b(s \times t,\, u)$, $s \otimes t \otimes u \mapsto b(s,\, t \times u)$ are non-zero. Similarly, $v \otimes w \otimes x \mapsto (v \times w,\, x)_V$, $v \otimes w \otimes x \mapsto (v,\, w \times x)_V$ are non-zero.

If either pair of functions were linearly independent, the other pair would be zero, contradicting $(*)$. So in each case, the two linear functions are scalar multiples of each other and the following six functions are non-zero scalar multiples of each other:

$$b(s\times t,\,u),\ b(t\times u,\,s),\ b(u\times s,\,t),\ b(t\times s,\,u),\ b(u\times t,\,s),\ b(s\times u,\,t).$$

We get a 1-dimensional representation of Σ_3, which is therefore trivial or the alternating representation and since $b(s\times t,\,u) = b(t\times s,\,u)$, it must be trivial. The same argument gives that the following six functions are non-zero scalar multiples of each other:

$$(v\times w,\,x)_V,\ (w\times x,\,v)_V,\ (x\times v,\,w)_V,\ (w\times v,\,x)_V,\ (x\times w,\,v)_V,\ (v\times x,\,w)_V,$$

and since $(v\times w,\,x)_V = -(w\times v,\,x)_V$, the representation of Σ_3 is the alternating representation. □

This finishes the second proof of Theorem 14.5. □

One can go much further with the properties of $\times$, b and $(\ ,\ )$ using for example the Jacobi identity in $L(E_8)$. Thus one can define

$$\begin{aligned} \circ : U\otimes U \to L(F_4), &\quad \text{by}\quad (a,\,s\circ t) = b([a,s],\,t),\\ \circ : V\otimes V \to L(G_2), &\quad \text{by}\quad (v',\,v\circ w) = ([v',v],\,w), \end{aligned}$$

both of which are anti-symmetric. The remaining two components of the Lie bracket $(U\otimes V)^2 \to L(E_8)$ can be given in terms of these products (we gave the component into $U\otimes V$ in Lemma 14.9). They are

$$\begin{aligned} (U\otimes V)^2 \to L(F_4), &\quad (s\otimes v)\otimes(t\otimes w) \mapsto (s\circ t)(v,w),\\ \to L(G_2), &\quad (s\otimes v)\otimes(t\otimes w) \mapsto b(s,t)v\circ w. \end{aligned}$$

The fact that these components can be given in this way follows from the invariance of the inner product in $L(E_8)$. Furthermore, we can describe the action of $s\circ t$ on U and of $v\circ w$ on V: we must have

$$[s\circ t, u] = \alpha(b(t,u)s - b(u,s)t) + \beta((t\times u)\times s - (u\times s)\times t), \qquad (**)$$

where α, β are scalars determined by various choices not made explicit. Considered as a function of u, $[s\circ t,\,u]$ is a derivation with respect to $\times$, that is,

$$[s\circ t,\,u\times u'\,] = [s\circ t,\,u]\times u' + u\times[s\circ t,\,u'\,],$$

so the right hand side of $(**)$ is a derivation and we have a non-trivial algebraic relation between b and $\times$. Similarly,

$$[v\circ w,\,x] = \gamma((w,x)v - (x,v)w) + \delta((v\times w)\times x + (w\times x)\times v + (x\times v)\times w),$$

where γ and δ are scalars, inexplicitly determined by various choices. As before, $[v \circ w, x]$ is a derivation with respect to x. We deduce from $(**)$ yet another proof that $\times$ is not zero. (If $\times$ were zero, we would have $\alpha \neq 0$ so that the linear functions of u in the α term would give us all the linear maps $U \to U$ in $L(SO(U))$ of dimension $26 \cdot 25/2 = 325$, and they cannot all come from $L(F_4)$.)

The algebra structure on U**.**

We now seek to show that the 26-dimensional representation U is isomorphic to an explicitly constructed algebra. (An *algebra*, over a field, is a vector space, A, together with a bilinear map $A \times A \to A$.) The difficulty is with the 2-dimensional part of U on which $Spin(8)$ acts trivially. Clearly $N_{F_4}(Spin(8))$ acts on this 2-dimensional subspace and hence $N_{F_4}(Spin(8))/Spin(8)$ acts on it, so $\Sigma_3 = Out(Spin(8))$ acts on it.

Proposition 14.11. *This 2-dimensional part is the (reduced) permutation representation of* Σ_3.

Proof. Let $SU(3) \times SU(3) \to F_4$ be as in Lemma 14.1. Then, as in the proof of Lemma 14.3, we have $U \mapsto \lambda^1 \otimes \lambda^1 + \lambda^2 \otimes \lambda^2 + 1 \otimes L(SU(3))$ over $SU(3) \times SU(3)$. The 2-dimensional subspace is $L(T)$ for the second $SU(3)$ (where the zero roots are). Our Σ_3 is the Weyl group $W(SU(3))$ of the second factor (see the proof of Theorem 14.2). The action of $W(SU(3))$ on its $L(T)$ is the reduced permutation representation. □

We have $\Sigma_2 = \dfrac{W(Spin(9))}{W(Spin(8))} \subset \dfrac{W(F_4)}{W(Spin(8))} = \Sigma_3$.

Corollary 14.12. *When the above* Σ_2 *acts on our 2-dimensional representation, the* $+1$ *eigenspace is the summand* 1 *in* $U \mapsto 1 + \lambda_9^1 + \Delta$ *over* $Spin(9)$ *and the* -1 *eigenspace is the summand* 1 *in* $\lambda_9^1 \mapsto 1 + \lambda_8^1$ *over* $Spin(8)$. □

The next task is to take the structure on U and see how it looks if we break it up according to the action of $Spin(8)$. First, a lemma.

Lemma 14.13. $Spin(8)$ *is transitive on pairs* (x, y) *of vectors* $x \in S^7 \subset \Delta^+$, $y \in S^7 \subset \Delta^-$.

Proof. We assume that we have a positive definite inner product on $\Delta^{\pm}$, invariant under $Spin(8)$. Certainly $Spin(8)$ is transitive on points $z \in S^7 \subset \lambda_8^1$. If we choose $z = (0, \ldots, 0, 1)$ the subgroup fixing z is $Spin(7)$ and Δ^+ restricts to Δ; $Spin(7)$ is transitive on points $x \in S^7 \subset \Delta$ by construction, so $Spin(8)$ is transitive on pairs (x, z), $x \in S^7 \subset \Delta^+$, $z \in S^7 \subset \lambda^1$. By triality, it is transitive on pairs (x, y). □

Recall that we have a triality map $f : \Delta^+ \otimes \Delta^- \otimes \lambda^1 \to \mathbf{R}$, invariant under $Spin(8)$ and $-1 \leq f(x \otimes y \otimes z) \leq 1$ when $||x|| = ||y|| = ||z|| = 1$; these bounds are attained. We thus find a product $\times : \Delta^+ \otimes \Delta^- \to \lambda^1$ such that $(x \times y, z) = f(x \otimes y \otimes z)$.

Lemma 14.14. *The above product satisfies* $||x \times y|| = ||x|| \, ||y||$.

Proof. First, let x, y, z attain the bounds, that is $||x|| = ||y|| = ||z|| = 1$ and $f(x \otimes y \otimes z) = (x \times y, z) = 1$. But $-1 \leq (x \times y, z') \leq 1$ for all z' with $||z'|| = 1$. Thus we must have $z = x \times y$, so for these x, y with $||x|| = 1 = ||y||$, we have $||x \times y|| = 1$. By Lemma 14.13, $Spin(8)$ is transitive on pairs (x, y) of vectors of unit length in Δ^+, Δ^-. So we have $||x \times y|| = 1$, if $||x|| = 1 = ||y||$. The general case follows by bilinearity. □

Corollary 14.15. $\mathbf{R}^8$ *admits the structure of a normed algebra with* 1.

Proof. Choose base points $x_0 \in S^7 \subset \Delta^+$, $y_0 \in S^7 \subset \Delta^-$, and identify

$$\begin{array}{lll} \Delta^- \text{ with } \lambda^1 = \mathbf{R}^8 & \text{by} & y \mapsto x_0 \times y, \\ \Delta^+ \text{ with } \lambda^1 & \text{by} & x \mapsto x \times y_0. \end{array}$$

With these identifications, $\times$ gives a normed product on $\mathbf{R}^8$ for which $x_0 \times y_0$ is a unit. □

For the moment we put this Corollary aside because we refuse to break symmetry by choosing base points x_0, y_0 in this way. We move to representations of $Spin(9)$.

Lemma 14.16. (i) *There is a trilinear map* $\Delta \otimes \Delta \otimes \lambda^1_9 \to \mathbf{R}$ *or* $\mathbf{C}$ *which is invariant under* $Spin(9)$ *and non-zero. It is unique up to a non-zero scalar and symmetric in the first two variables.*

(ii) *Suppose given decompositions* $\Delta \xleftarrow{\cong} \Delta^+ + \Delta^-$, $\lambda^1_9 \xleftarrow{\cong} \lambda^1_8 + 1$ *over* $Spin(8)$ *preserving inner products. Then, up to a scalar factor, the trilinear map becomes*

$$(a^+ + a^-) \otimes (b^+ + b^-) \otimes (z + \zeta) \mapsto f(a^+ \otimes b^- \otimes z) + f(b^+ \otimes a^- \otimes z) \pm \zeta((a^+, b^+) - (a^-, b^-)),$$

where the sign is fixed for any set of choices, but can be changed by changing the sign of any of $\Delta^+ \to \Delta$, $\Delta^- \to \Delta$, $\lambda^1_8 \to \lambda^1_9$, $1 \to \lambda^1_9$ *or* f.

Proof. (i) By transposing, it is equally well to consider $\Delta \otimes \Delta \to \lambda^1_9$. But we have $\sigma^2(\Delta) = \lambda^9 + \lambda^1 + 1$ and $\lambda^2(\Delta) = \lambda^3_9 + \lambda^2_9$, by Theorem 4.6, so up to a scalar multiple, there is just one such linear map and it is symmetric.

(ii) Because the map is invariant under $Spin(8)$, the map in part (i), regarded as a map $\Delta \otimes \Delta \to \lambda^1_9$, must decompose as

$$(a^+ + a^-) \otimes (b^+ + b^-) \mapsto \big(\alpha(a^+ \times b^-) + \beta(b^+ \times a^-), \gamma(a^+, b^+) + \delta(a^-, b^-)\big),$$

where α, β, γ, δ are scalars and $\times : \Delta^+ \otimes \Delta^- \to \lambda_8^1$ is as in Lemma 14.14. By symmetry $\alpha = \beta$.

Consider the effect of an element $g \in Spin(9)$, covering $\begin{pmatrix} A & 0 \\ 0 & -1 \end{pmatrix}$, where $\det A = -1$. This element g acts on Δ and interchanges Δ^+, Δ^- preserving their inner products. Suppose it carries a^+ of length 1 to a^- of length 1. Then we have

$$\begin{array}{ccc} (a^+ + 0) \otimes (a^+ + 0) & \longrightarrow & (0, \gamma) \\ {\scriptstyle g\otimes g}\big\downarrow & & \big\downarrow{\scriptstyle g} \\ & & (0, -\gamma) \\ (0 + a^-) \otimes (0 + a^-) & \longrightarrow & (0, \delta) \end{array} .$$

Thus $\gamma = -\delta$ and we are down to $\big(\alpha a^+ \times b^- + \alpha b^+ \times a^-,\ \gamma(a^+, b^+) - \gamma(a^-, b^-)\big)$.

Now consider the 1-parameter subgroup joining 1 to such an element in $Spin(9)$, say $g_\theta = \cos\theta + (e_8 e_9)\sin\theta$. Acting on $(a^+, 0) \in \Delta$ we get, in Δ, a path of elements $g_\theta(a^+, 0) = (\cos\theta)a^+ + (\sin\theta)a^-$, $a^- \in \Delta^-$, $||a^\pm|| = 1$. Under our map $\Delta \otimes \Delta \to \lambda_9^1$ we have

$$\begin{aligned} &\big((\cos\theta)a^+ + (\sin\theta)a^-\big) \otimes \big((\cos\theta)a^+ + (\sin\theta)a^-\big) \\ &\qquad \mapsto \big((2\alpha\cos\theta\sin\theta)a^+ \times a^-,\ \gamma(\cos^2\theta - \sin^2\theta)\big), \end{aligned}$$

which must have constant length, because it is obtained by $g_\theta \in Spin(9)$ acting on a fixed vector. But $||a^+ \times a^-|| = 1$ so that the square of the length of the right hand side is $\alpha^2 \sin^2 2\theta + \gamma^2 \cos^2 2\theta$. But this must be γ^2. Thus $\alpha^2 = \gamma^2$, so $\gamma = \pm\alpha$. □

We are now ready to give the algebra structure on U. Let V_1, V_2, V_3 be Δ^+, Δ^-, λ_8^1 in some order and $V_i \to U$ some embedding, invariant under $Spin(8)$ and preserving the inner product. Take $\mathbf{R}^3$ and let Σ_3 act on it by permuting the basis vectors e_1, e_2, e_3 as it permutes V_1, V_2, V_3. Take the subspace $S \subset \mathbf{R}^3$ defined by $x_1 + x_2 + x_3 = 0$; this is the reduced representation. Define an inner product on $\mathbf{R}^3$ by $(x, y) = \frac{1}{2}(x_1y_1 + x_2y_2 + x_3y_3)$ and let $S \to U$ be the inclusion onto the two dimensional subspace invariant under Σ_3 and preserving the inner product. We also need the vectors which change sign under a 2-cycle: the vectors $e_1 - e_2$, $e_2 - e_3$, $e_3 - e_1$ will be convenient.

Lemma 14.17. *With respect to this decomposition, the invariant trilinear function $U \otimes U \otimes U \to \mathbf{R}$ or $\mathbf{C}$ has (up to a scalar factor) the following components:*

(i) *On $V_i \otimes V_j \otimes V_k$ with i, j, k all different, permute the variables to get to $\Delta^+ \otimes \Delta^- \otimes \lambda^1$ and use f.*

(ii) *On $S \otimes V_i \otimes V_i$ take $x \otimes y \otimes z$ to $\pm x_i(y, z)$ and similarly on $V_i \otimes S \otimes V_i$ and $V_i \otimes V_i \otimes S$.*

(iii) *On $S \otimes S \otimes S$ take $x \otimes y \otimes z$ to $\mp(x_1y_1z_1 + x_2y_2z_2 + x_3y_3z_3)$.*

(iv) *On all other summands,* 0.

The signs $\pm$, $\mp$ *are opposite and fixed for any set of choices but can be changed (by changing the sign of* $S \hookrightarrow U$*).*

Proof. The components of the invariant function $U \otimes U \otimes U \to \mathbf{R}$ or $\mathbf{C}$ are certainly invariant under $Spin(8)$. From this it follows that they are 0 except in the stated cases. We have $U \hookleftarrow S + \begin{cases} \Delta^+ + \Delta^- + \lambda^1, \\ V_1 + V_2 + V_3 \end{cases}$ where, of course, U is the 26-dimensional representation of F_4.

(i) On $V_i \otimes V_j \otimes V_k$ we get $f : \Delta^+ \otimes \Delta^- \otimes \lambda^1 \to \mathbf{R}$ after permutation of the variables (up to non-zero scalar factor). Clearly all repeated term products vanish, and both the previous proofs that $U^3 \to \mathbf{R}$ is non-zero show that $(\Delta^+ + \Delta^- + \lambda^1)^3 \to \mathbf{R}$ is non-zero. We may assume without loss of generality that the scalar is 1.

(ii) On $S \otimes V_i \otimes V_i$, $x \otimes y \otimes z$ goes to $\ell_i(x)(y, z)$ for some linear $\ell_i : S \to \mathbf{R}$ or $\mathbf{C}$, which must be invariant under the 2-cycle in Σ_3 which interchanges V_j and V_k where $i \neq j \neq k \neq i$, so ℓ_i must be a scalar multiple of the function $x \mapsto x_i$. Thus we get $x \otimes y \otimes z \mapsto \alpha_i x_i(y, z)$ for some scalar α_i. Because of the Σ_3-symmetry, we must have $\alpha_1 = \alpha_2 = \alpha_3 = \alpha$, say.

(iii) Similarly, representation theory shows that on $S \otimes S \otimes S$ there is only one Σ_3-invariant trilinear function up to a scalar multiple. The function given is such a function so we have $x \otimes y \otimes z \mapsto \beta \sum_i x_i y_i z_i$ for some scalar β.

We proceed to determine α, β. Consider $Spin(9) \subset F_4$ and first determine α. The restriction of our invariant map to $\Delta \otimes \Delta \otimes \lambda_9^1$ is invariant under $Spin(9)$ and Lemma 14.16 applies to it. We see that the components agree with those here. Let us write e_+, e_- for the e_i corresponding to Δ^+, Δ^-, so that $e_\pm \in \Delta^\pm, e_+ - e_-, 2e_0 - e_+ - e_- \in S$. Thus we have:

$$\begin{aligned} (a^+ + a^-) \otimes (b^+ + b^-) \otimes (e_+ - e_-) &\mapsto \pm\left((a^+, b^+) - (a^-, b^-)\right), \text{ by Lemma 14.16,} \\ &\mapsto \alpha\left((a^+, b^+) - (a^-, b^-)\right), \text{ by the above.} \end{aligned}$$

Thus $\alpha = \pm 1$.

To determine β, note that the restriction of our invariant map to $\lambda_9^1 \otimes \lambda_9^1 \otimes 1$ is also invariant under $Spin(9)$. Now, on $\lambda_9^1 \otimes \lambda_9^1 \otimes 1$ we have

$$v \otimes v' \otimes (2e_0 - e_+ - e_-) \mapsto 2\alpha(v, v'), \qquad v, v' \in \lambda_8^1.$$

By invariance under $Spin(9)$, this must also hold for $v, v' \in \lambda_9^1$. In particular,

$$(e_+ - e_-) \otimes (e_+ - e_-) \otimes (2e_0 - e_+ - e_-) \mapsto 2\alpha,$$

where $e_+ - e_- \in 1 \subset \lambda_9^1 = 1 + \lambda_8^1$. But by the above, this maps to -2β. Thus $\beta = -\alpha$. $\square$

We now create a 27-dimensional algebra with a unit since this is the usual construction (rather than a 26-dimensional representation U, without unit). We restate Lemma 14.17 in the new notation.

Theorem 14.18. *Take the 27-dimensional space* $\mathbf{R}^3 + V_1 + V_2 + V_3$ *and place on it the following structure:*

(i) *a linear form defined on* $\mathbf{R}^3$ *by* $\ell(x) = x_1 + x_2 + x_3$ *and zero on all other summands;*

(ii) *a symmetric bilinear form* b *defined on* $\mathbf{R}^3 \otimes \mathbf{R}^3$ *to be the usual inner product, on* $V_i \otimes V_i$ *twice the usual inner product and zero elsewhere;*

(iii) *a symmetric form* t *defined on* $V_i \otimes V_j \otimes V_k$ $(i \neq j \neq k \neq i)$ *by permuting the variables and then using* $f : \Delta^+ \otimes \Delta^- \otimes \lambda_8^1 \to \mathbf{R}$,

$$\text{on } \mathbf{R}^3 \otimes V_i \otimes V_j, \qquad e_k \otimes v \otimes v' \mapsto \begin{cases} (v, v') & \text{if } j = i \neq k \\ 0 & \text{if } j \neq i; \end{cases}$$

$$\text{on } \mathbf{R}^3 \otimes \mathbf{R}^3 \otimes \mathbf{R}^3, \qquad e_i \otimes e_j \otimes e_k \mapsto \begin{cases} 1 & \text{if } i = j = k \\ 0 & \text{otherwise.} \end{cases}$$

On all other summands t *is zero.*

Then, the subspace $\ell = 0$ *is isomorphic to the 26-dimensional representation* U *of* F_4 *carrying* b *and* t *onto the two structure maps up to a scalar for each. See Theorem 16.1.* (We arrange to have $\beta = 1$, $\alpha = -1$ by, for example, choice of $S \to U$.)

Proof. In the proof of Lemma 14.17, we take $\alpha = -1$ and $x \otimes v \otimes v' \mapsto \ell_i(x)(v, v')$, $\ell_i(x) = -x_i$. But in the statement of the theorem, $\ell_i(x) = x_j + x_k$. This equals $-x_i$ on S, since $x_1 + x_2 + x_3 = 0$ there. □

To make the algebra product quite explicit, let $J = \mathbf{R}^3 + V_1 + V_2 + V_3$. Then the product $\circ : J \times J \to J$ is defined by $b(x \circ y, z) = t(x, y, z)$. Compare Theorem 16.1 and the definitions preceding it.

We have seen that the crucial part of the structure of U is the trilinear map f and that f is connected with the existence of a normed algebra in dimension 8. In the next chapter we review the construction of this normed algebra.

Chapter 15

The Cayley numbers

What is at the root of all the exceptional phenomena we have to study: the trilinear maps of Proposition 5.7, Lemma 14.16 and the skew symmetric product of Corollary 14.7? We will see that F_4 and hence the other exceptional groups are intimately related to the algebra of Cayley numbers. In particular (Theorem 15.14), we give a construction of the triality of Proposition 5.7, using the Cayley numbers. We also prepare results such as Lemma 15.9 and Corollary 15.12 for use in Chapter 16 to construct F_4 and its 27-dimensional representation, independently of the previous Chapters, see Theorem 16.1. A reference for this Chapter is [11]; see also [Fr].

Theorem 15.1. *There is a unique normed algebra of dimension* 8 *over* $\mathbf{R}$. (A *normed* algebra is one for which $||xy|| = ||x||\,||y||$.)

Proof. If we just wanted to prove the existence part, we could quote Corollary 14.15, but as we want uniqueness also, we may as well prove that first, as it will lead to the construction of the Cayley numbers. Suppose we have a linear map $\mu : U \otimes V \to W$ with $||\mu(u \otimes v)|| = ||u||\,||v||$ and U, V, W vector spaces over $\mathbf{R}$ with $\dim V = \dim W$, $\dim U > 0$ and $||u_0|| = 1$ for some $u_0 \in U$. We will write $\mu(u \otimes v)$ as uv.

The map $V \to W$, $v \mapsto uv$, is an isomorphism, so without loss of generality, we can assume $U = \mathbf{R}^p$, $V = W = \mathbf{R}^n$ (with the usual Euclidean norm). Identify u_0 with $1 \in \mathbf{R}^p$ since it acts as a unit. Assume that $p \geq 2$ and 1, i are orthonormal in $\mathbf{R}^p$.

Lemma 15.2. *For all* $x \in \mathbf{R}^n$, $i(ix) = -x$.

Proof. $x \mapsto ix$ is an isometry of $\mathbf{R}^n$ so by choice of an orthonormal basis, it can be written

$$\begin{pmatrix} \cos\theta_1 & -\sin\theta_1 & & & & \\ \sin\theta_1 & \cos\theta_1 & & & & \\ & & \cos\theta_2 & -\sin\theta_2 & & \\ & & \sin\theta_2 & \cos\theta_2 & & \\ & & & & \ddots & \\ & & & & & \pm 1 \end{pmatrix}$$

We also have

$$||(a+bi)x||^2 = ||a+bi||^2||x||^2$$

for all $a, b \in \mathbf{R}$, so

$$a^2(x,x) + 2ab(x,ix) + b^2(ix,ix) = (a^2+b^2)(x,x)$$

and thus $(x, ix) = 0$, that is, each $\theta_i = \pm\pi/2$ and there are no blocks of ± 1. Hence the square of the isometry is -1. □

Let $\mathbf{R}^{p-1} \subset \mathbf{R}^p$ be the orthogonal complement of $\langle 1 \rangle$, the subspace generated by 1.

Lemma 15.3. *$Cl(\mathbf{R}^{p-1})$ acts on $\mathbf{R}^n$.*

Proof. For all $u \in \mathbf{R}^{p-1}$, we have a linear map $\mathbf{R}^n \to \mathbf{R}^n$, $x \mapsto ux$, giving a linear map $\mathbf{R}^{p-1} \to Hom(\mathbf{R}^n, \mathbf{R}^n)$, which extends to $T(\mathbf{R}^{p-1}) \to Hom(\mathbf{R}^n, \mathbf{R}^n)$, where $T(\mathbf{R}^{p-1})$ is the tensor algebra. We check that $u^2 \mapsto -(u,u)1$: this is clear if $u = 0$; otherwise it is sufficient to check the case $||u|| = 1$, which we do from Lemma 15.2. □

Corollary 15.4. *If $p > 2$, and $1, i, j$ are orthonormal in $\mathbf{R}^p$, then $i(jx) = -j(ix)$ for all $x \in \mathbf{R}^n$.*

Proof. By Lemma 15.3, $Cl(\mathbf{R}^2) = \mathbf{H}$ acts on $\mathbf{R}^n$ and in $\mathbf{H}$, $ij = -ji$. □

Of course, this is the line of thought which shows that a normed product $\mathbf{R}^n \otimes \mathbf{R}^n \to \mathbf{R}^n$ exists only if $n = 1, 2, 4$ or 8. This is because $\mathbf{R}^n$ becomes a module over $Cl(\mathbf{R}^{n-1})$ and so n is divisible by a large power of 2 depending on $n-1$, which reduces things to the cases mentioned.

Assume that $p = n > 2$; then by choosing $v_0 \in V$ with $||v_0|| = 1$, we can identify U with W and so arrange that $\mathbf{R}^n$ has the structure of a normed algebra with a two sided unit 1. We name this algebra A.

Corollary 15.5. *If $1, i$ are orthonormal in A then for all $a \in A$, $i(ia) = -a$ and in particular $i^2 = -1$. If $1, i, j$ are orthonormal in A then $i(ja) = -j(ia)$ and in particular $ij = -ji$.*

Proof. See Lemma 15.2 and Corollary 15.4. □

Lemma 15.6. *Suppose that* $1, i, j$ *are orthonormal in* A. *Then* $1, i, j, ij$ *are orthonormal in* A *and they form a basis for a subalgebra isomorphic to* $\mathbf{H}$. *In particular, the subalgebra is associative.*

Proof. (We must avoid using associativity, which we do not have.) Since $1, i, j$ are orthonormal and $x \mapsto ix$ is an isometry, i, $i^2 = -1$, ij are orthonormal. Similarly $x \to xj$ is an isometry, 1, i orthonormal, so j, ij are orthonormal. Thus $1, i, j, ij$ are orthonormal. To avoid confusion we will write $k = ij$ for the product in A. Let $\mathbf{R}^3$ be the subspace spanned by i, j, k. By Lemma 15.3, we have an action of $Cl(\mathbf{R}^3)$ on A. Define $\theta : Cl(\mathbf{R}^3) \to A$ by $c \mapsto c \cdot 1$ and consider $\theta(k(ij)) = (k(ij)) \cdot 1 = k(ij) = (ij)(ij) = -1$, by Corollary 15.5. Thus $\theta(kij + 1) = 0$, so θ factors to a map $\tilde{\theta} : Cl(\mathbf{R}^3)/(Cl(\mathbf{R}^3))(kij + 1) \to A$. The left ideal $(Cl(\mathbf{R}^3))(kij + 1)$ is actually a two-sided ideal since kij is central in $Cl(\mathbf{R}^3)$ and so $Cl(\mathbf{R}^3)/(Cl(\mathbf{R}^3))(kij + 1)$ is an algebra: in fact it is $\mathbf{H}$. In particular,

$$Cl(\mathbf{R}^3)/(Cl(\mathbf{R}^3))(kij + 1) = \mathbf{R} + \mathbf{R}^3$$

where $\mathbf{R}$ is generated by 1 and $\mathbf{R}^3$ by i, j, k. Thus $\tilde{\theta}$ is an isomorphism onto the subspace of A with basis $1, i, j, ij$. (Note that for a subalgebra of A of dimension greater than 3, the argument would fail here since not all basis elements can be written so simply.)

Claim. $\tilde{\theta}$ *preserves products:* $\tilde{\theta}(uv) = \tilde{\theta}(u)\tilde{\theta}(v)$.

Proof. Clearly the Claim is true if either u or v is 1 (as $\tilde{\theta}(1) = 1$), so it is sufficient to take $u, v \in \mathbf{R}^3$. Then

$$\begin{aligned} \tilde{\theta}(uv) &= (uv) \cdot 1 &&= uv, \\ \tilde{\theta}(u)\tilde{\theta}(v) &= (u \cdot 1)(v \cdot 1) &&= uv. \end{aligned}$$ □

That finishes the proof of the Lemma. □

Lemma 15.7. *Suppose* $1, i, j, ij, \ell$ *are orthonormal in* A. *Then* $i(j\ell) = -(ij)\ell$, *so that if this is not zero, we do not have associativity in* A.

Proof.

$$\begin{aligned} i(j\ell) &= -i(\ell j) && \text{by Lemma 15.5 as } 1, j, \ell \text{ are orthonormal,} \\ &= \ell(ij) && \text{taking } a = j \text{ and } 1, i, \ell \text{ the orthonormal set in Lemma 15.5,} \\ &= -(ij)\ell && \text{taking } 1, ij, \ell \text{ as the orthonormal set in Lemma 15.5.} \end{aligned}$$ □

Theorem 15.8. *Suppose* A *is a normed algebra with* 1 *of dimension* 8 *over* $\mathbf{R}$ *with* $1, i, j, ij, \ell$ *orthonormal in* A, *and* $\mathbf{H}$ *is the subalgebra corresponding* *to* 1, i, j, ij *as in Lemma* 15.6. *Then we have an isometry* $\mathbf{H} \oplus \mathbf{H} \to A$, $(a, b) \mapsto a + b\ell$ *and the product in* A *is given by* $(a, b)(c, d) = (ac - \bar{d}b, da + b\bar{c})$.

Proof. ℓ is orthogonal to $\mathbf{H}$ and $a \mapsto ia$, $a \mapsto ja$, $a \mapsto (ij)a$ are isometries of A taking $\mathbf{H}$ to $\mathbf{H}$. Thus $i\ell$, $j\ell$, $(ij)\ell$ are orthogonal to $\mathbf{H}$. Also $a \mapsto a\ell$ is an isometry and 1, i, j, ij orthonormal, so ℓ, $i\ell$, $j\ell$, $(ij)\ell$ are orthonormal. Thus they are an orthonormal basis for the orthogonal complement of $\mathbf{H}$. We see that 1, i, j, ij, ℓ, $i\ell$, $j\ell$, $(ij)\ell$ are an orthonormal basis for A and $\mathbf{H} \oplus \mathbf{H} \to A$, $(a, b) \mapsto a + b\ell$ is an isometry. It remains to work out the products.

Claim 1. *For all* $a, d \in A$, $a(d\ell) = (da)\ell$.

Proof. It is sufficient to prove the Claim for $a, d \in \{1, i, j, ij\}$ and it is trivial for a or $d = 1$. If $a = d$ then $a(a\ell) = -\ell$ and $(aa)\ell = -\ell$ by Lemma 15.5. If $a \neq d$, then

$$\begin{aligned} a(d\ell) &= -(ad)\ell && \text{by Lemma 15.7,} \\ &= (da)\ell && \text{by Lemma 15.5.} \end{aligned}$$

□

Claim 2. For $c, d \in \mathbf{H}$, we have $(b\ell)c = (b\bar{c})\ell$.

Proof. The Claim is clear if $c = 1$, so assume $b \in \{1, i, j, ij\}$, $c \in \{i, j, ij\}$. Then

$$\begin{aligned} (b\ell)c &= -\,c(b\ell) && \text{by Lemma 15.5 since } b\ell,\ c,\ 1 \text{ are orthonormal,} \\ &= -\,(bc)\ell && \text{by Claim 1,} \\ &= (b\bar{c})\ell. \end{aligned}$$

□

Finally,

$$(b\ell)(\bar{d}b) = (b\overline{(\bar{d}b)})\ell = (b(\bar{b}d))\ell = d\ell$$

for $b \in \{1, i, j, ij\}$. Thus

$$(b\ell)(d\ell) = (b\ell)((b\ell)(\bar{d}b)) = -\,\bar{d}b$$

by repeated use of Corollary 15.5.

From these Claims, we conclude that $(a, b)(c, d) = (ac - \bar{d}b, da + b\bar{c})$. This finishes the proof of Theorem 15.8 and the uniqueness statement in Theorem 15.1.

□

Now we prove the existence part of Theorem 15.1 by showing that $\mathbf{H} \oplus \mathbf{H}$ *is* a normed algebra with the expected properties.

Definition. $A = \mathbf{H} \oplus \mathbf{H}$ *with the product* $(a, b)(c, d) = (ac - \bar{d}b, da + b\bar{c})$ *and* $(1, 0)$ *as unit is called the* algebra of Cayley numbers $\mathbf{O}$. *The* conjugate *of* (a, b) *is* $\overline{(a, b)} = (\bar{a}, -b)$. *The* $+1$ *eigenspace of conjugation is* $\langle (1, 0) \rangle_{\mathbf{R}}$ *of* *dimension* 1 *and called the set of* real parts, *and the* -1 *eigenspace of dimension* 7 *is called the set of* pure imaginary *Cayley numbers.*

Note. The construction of $\mathbf{O}$ is a generalisation of $\mathbf{C} = \mathbf{R} \oplus \mathbf{R}$ and $\mathbf{H} = \mathbf{C} \oplus \mathbf{C}$. The letter $\mathbf{O}$ comes from "octonians".

Lemma 15.9. *For all* $x, y \in \mathbf{O}$,

(i) $\bar{\bar{x}} = x$,

(ii) $\overline{xy} = \bar{y}\bar{x}$,

(iii) $Re(xy) = Re(yx)$; *for pure imaginary Cayley numbers we have* $Im(xy) = -Im(yx)$,

(iv) $x\bar{x} = ||x||^2 \cdot 1 = \bar{x}x$,

(v) $(x, y) = (\bar{x}, \bar{y}) = Re(x\bar{y}) = Re(\bar{y}x)$.

Proof. (i) Clear.

(ii)
$$\begin{aligned} \bar{y}\bar{x} &= (\bar{c}, -d)(\bar{a}, -b) = (\bar{c}\bar{a} - \bar{b}d, -b\bar{c} - da), \\ \overline{xy} &= \overline{(a,b)(c,d)} \quad = (\bar{c}\bar{a} - \bar{b}d, -da - b\bar{c}). \end{aligned}$$

(iii) For pure imaginary Cayley numbers x, y we have $\overline{xy} = \bar{y}\bar{x} = yx$, so $Re(yx) = Re(xy)$ (from which it follows that $Re(xy) = Re(yx)$ for all Cayley numbers x, y) and $Im(xy) = -Im(yx)$.

(iv) $$x\bar{x} = (a,b)(\bar{a}, -b) = (a\bar{a} + \bar{b}b, -ba + ba) = (||a||^2 + ||b||^2)(1, 0).$$

(v) $(x, y) = (\bar{x}, \bar{y})$ because conjugation is an isometry. Take the expression for $||x||^2$ and polarise it. □

We next determine when associativity holds.

Lemma 15.10. *For all* $x, y \in \mathbf{O}$, *we have*

$$\bar{x}(xy) = (\bar{x}x)y \quad \text{and} \quad (xy)\bar{y} = x(y\bar{y}).$$

Proof. Letting $x = (a, b)$, $y = (c, d)$, we have

$$\begin{aligned} \overline{(a,b)}((a,b)(c,d)) &= (\bar{a}, -b)(ac - \bar{d}b,\ da + b\bar{c}) \\ &= (\bar{a}ac - \bar{a}\bar{d}b + \bar{a}\bar{d}b + c\bar{b}b,\ da\bar{a} + b\bar{c}\bar{a} - b\bar{c}\bar{a} + b\bar{b}d) \\ &= (||a||^2 + ||b||^2)(c, d) \\ &= (\bar{x}x)y. \end{aligned}$$

For the second part, putting $\bar{y}$ for x and $\bar{x}$ for y gives $y(\bar{y}\bar{x}) = (y\bar{y})\bar{x}$ which on taking conjugates leads to $(xy)\bar{y} = x(y\bar{y})$. □

We can express these results in terms of *alternative algebras.*

Definition. *In an algebra A, which need not be associative, the* associator *of $x, y, z \in A$ is $[x,y,z] = (xy)z - x(yz)$.*

The algebra A is alternative *if $[x,y,z]$ is an alternating function of x, y, z.*

Theorem 15.11. *The Cayley numbers are alternative. In detail, the associator $[x,y,z]$*

(i) *is pure imaginary,*

(ii) *changes sign if you conjugate any variable,*

(iii) *is an alternating function of the three variables.*

Proof. (ii) Clearly $[x,y,z] = 0$ if $x = 1$, $y = 1$ or $z = 1$, so in proving (ii), we can assume the variable to be conjugated is pure imaginary. Thus for example $[\bar{x},y,z] = [-x,y,z] = -[x,y,z]$.

(iii) By Lemma 15.10, $[\bar{x},x,y] = 0 = [x,y,\bar{y}]$, so by (ii), $[x,x,y] = 0 = [x,y,y]$, and polarisation gives us $[w+x,\ w+x,\ y] = 0 = [x,\ y+z,\ y+z]$ which lead to $[w,x,y] + [x,w,y] = 0$, $[x,y,z] + [x,z,y] = 0$. That is, the associator is alternating if you change its first two or last two variables. These transpositions generate Σ_3. For example, $[x,y,z] = -[y,x,z] = [y,z,x] = -[z,y,x]$.

(i)
$$\begin{aligned}
\overline{[x,y,z]} &= \overline{(xy)z - x(yz)} \\
&= \bar{z}(\bar{y}\bar{x}) - (\bar{z}\bar{y})\bar{x} \\
&= -[\bar{z},\bar{y},\bar{x}] \\
&= [z,y,x] && \text{by (ii),} \\
&= -[x,y,z] && \text{by (iii).}
\end{aligned}$$
□

Corollary 15.12. (i) *For Cayley numbers x, y, z the real parts of $(xy)z$, $(yz)x$, $(zx)y$, $x(yz)$, $y(zx)$, $z(xy)$ are equal.*

(ii) *For pure imaginary Cayley numbers x, y, z, we have $(xy,z) = (x,yz)$ and it is an alternating function of its three variables.*

(Compare part (ii) with Lemma 14.10.)

Proof. From Theorem 15.11 (i), we have $Re[x,y,z] = 0$, that is, $Re(x(yz) - (xy)z) = 0$, so $Re(x(yz)) = Re((xy)z)$. But by Lemma 15.9, $Re(x(yz)) = Re((yz)x)$ and so on.

(ii)
$$\begin{aligned}
(x,yz) &= Re((yz)\bar{x}) && \text{by Lemma 15.9 (v),} \\
&= -Re((yz)x) && \text{since } x \text{ is imaginary,} \\
&= -Re(z(xy)) && \text{by (i),} \\
&= Re(\bar{z}(xy))
\end{aligned}$$

$= (xy,\, z).$

Also $(yx, z) = -(xy, z)$ by Lemma 15.9 (v). □

Corollary 15.13. $||x||\,||y|| = ||xy||$.

Proof.
$$\begin{aligned} ||xy||^2 &= Re((\overline{xy})(xy)) \\ &= Re((\bar{y}\bar{x})(xy)) \\ &= Re(\bar{y}(\bar{x}(xy))) \quad \text{by the previous corollary,} \\ &= Re(\bar{y}((\bar{x}x)y)) \quad \text{by Lemma 15.10,} \\ &= ||x||^2 Re(\bar{y}y) \\ &= ||x||^2||y||^2. \end{aligned}$$
□

We have thus shown that the Cayley numbers form a normed algebra. This finishes the proof of Theorem 15.1. □

Connection between the Cayley numbers and Lie groups.

Recall from Proposition 5.7 the trilinear function $f : \Delta^+ \otimes \Delta^- \otimes \lambda^1 \to \mathbf{R}$, invariant under *Spin*(8). Make fixed choices of unit vectors x_0, y_0, z_0 in Δ^+, Δ^-, λ^1 such that $f(x_0 \otimes y_0 \otimes z_0) = 1$.

Theorem 15.14. *With this choice of* (x_0, y_0, z_0), *we can identify* Δ^+, Δ^-, λ^1 *with the space* $\mathbf{O}$ *of Cayley numbers so as to preserve the inner product and map* x_0, y_0, z_0 *onto* 1 *and* f *onto the function* $Re((xy)z) = Re((yz)x)$ *etc. in Corollary* 15.12 (i).

Proof. In Lemma 14.14, we transposed f to give a product $\times : \Delta^+ \otimes \Delta^- \to \lambda^1$ satisfying $f(x \otimes y \otimes z) = (x \times y, z)$ and $||x \times y|| = ||x||\,||y||$. As we have seen, choice of x_0, y_0, z_0 allows one to identify the three vector spaces and get a normed algebra with 1 corresponding to x_0, y_0, z_0. This has to be isomorphic to the Cayley numbers. The isomorphism preserves the inner product and the product, and maps $f(x \otimes y \otimes z)$ onto $(xy, z) = Re((xy)\bar{z})$. If we alter the identification of λ^1 with Cayley numbers by $z \mapsto \bar{z}$, we still have z_0 corresponding to 1. □

Lemma 15.15. *Suppose given a triple of linear maps* $\theta : \Delta^+ \to \Delta^+$, $\varphi : \Delta^- \to \Delta^-$, $\psi : \lambda^1 \to \lambda^1$ *which preserves inner products and we have*

$$f(\theta x \otimes \varphi y \otimes \psi z) = f(x \otimes y \otimes z).$$

Then there exists a unique $g \in Spin(8)$ *acting as* θ *on* Δ^+, φ *on* Δ^- *and* ψ *on* λ^1.

Proof. The uniqueness is immediate. For the existence, we can suppose without loss of generality that $\theta = 1$ for we certainly have a $g \in Spin(8)$ cancelling θ. When we choose a base point $x_0 \in \Delta^+$, we get an action of $Cl(\mathbf{R}^7)$

on λ^1 where $\mathbf{R}^7$ is the orthogonal complement of x_0, and λ^1 has to be an irreducible representation of $Cl(\mathbf{R}^7)$. But ψ is a map of $Cl(\mathbf{R}^7)$-modules, so it has to be multiplication by a real scalar, which has to be ± 1, since the inner product is preserved. Similarly $\varphi = \pm 1$ and these signs are the same as f is preserved (or use the identification of Δ^- and λ^1 under multiplication by x_0). If both signs are -1, multiply g by $g' \in Spin(8)$ such that g' is 1 on Δ^+ and -1 on Δ^- and λ^1. □

Finally, as with the other exceptional groups, we identify G_2 as a group of automorphisms.

Theorem 15.16. $G_2 \cong Aut_{\mathbf{R}}(\mathbf{O})$.

Proof. By the proof of Theorem 15.14, $Aut_{\mathbf{R}}(\mathbf{O})$ is the group of linear maps $\theta : \Delta^+ \to \Delta^+$, $\varphi : \Delta^- \to \Delta^-$, $\psi : \lambda^1 \to \lambda^1$ preserving x_0, y_0, z_0, inner products and f. By Lemma 15.15, this is the subgroup of $Spin(8)$ preserving x_0, y_0, z_0. It is sufficient to preserve two of them, say x_0 and z_0, since they determine the third. The subgroup preserving z_0 is $Spin(7)$ and the restriction of Δ^+ to $Spin(7)$ is Δ, so we come down to the subgroup of $Spin(7)$ preserving $x_0 \in \Delta$. This is G_2 by definition (Theorem 5.5). □

Chapter 16

Direct treatment of F_4, II: Jordan algebras

Definition and properties of the exceptional Jordan algebra J.

Here we complete the direct treatment of F_4 begun in Chapter 14, by, for example, identifying the 27-dimensional representation of Theorem 14.18 with an algebra J and F_4 as the automorphisms of J, (Theorems 16.1 and 16.7). In the last section we use these results on J to define the Cayley plane and derive its basic properties. There is a brief appendix on the definition of *Jordan algebras*, special and exceptional.

The *exceptional Jordan algebra* J is usually defined as the set of 3×3 matrices $A = [a_{ij}]$, $a_{ij} \in \mathbf{O}$ which are Hermitian: $\bar{a}_{ij} = a_{ji}$ (so $a_{ii} \in \mathbf{R}$). The real dimension of J is 27 and a typical element has the form $A = \begin{pmatrix} \lambda_1 & x_3 & \bar{x}_2 \\ \bar{x}_3 & \lambda_2 & x_1 \\ x_2 & \bar{x}_1 & \lambda_3 \end{pmatrix}$, $\lambda_i \in \mathbf{R}$, $x_i \in \mathbf{O}$.

On this set J we have the following structure:

(i) a product $\circ : J \otimes J \to J$, $A \otimes B \mapsto A \circ B = \frac{1}{2}(AB + BA)$; (We check that $A \circ B$ is Hermitian:

$$\begin{aligned}(\overline{A \circ B})_{ik} &= \text{conjugate of } \tfrac{1}{2}\sum_j (a_{ij}b_{jk} + b_{ij}a_{jk}) \\ &= \tfrac{1}{2}\sum_j (\bar{a}_{jk}\bar{b}_{ij} + \bar{b}_{jk}\bar{a}_{ij}) \\ &= \tfrac{1}{2}\sum_j (a_{kj}b_{ji} + b_{kj}a_{ji}) \\ &= (A \circ B)_{ki}.)\end{aligned}$$

(ii) a linear map $\ell : J \to \mathbf{R}$ defined by $A \mapsto \lambda_1 + \lambda_2 + \lambda_3$; (Note that since $\ell(A) = \frac{1}{9}Tr(B \mapsto A \circ B)$, this is determined by the product.)

(iii) a bilinear map $b : J \otimes J \to \mathbf{R}$, $A \otimes B \mapsto \ell(A \circ B)$;

(iv) a trilinear map $t : J \otimes J \otimes J \to \mathbf{R}$, $A \otimes B \otimes C \mapsto b(A \circ B, C)$. (This is symmetric, but not obviously.)

Notes. (i) $A \circ B = B \circ A$.

(ii) $diag(1,1,1)$ is a unit.

(iii) $A \circ A = A^2$.

(iv) ℓ, b, t are determined by the product and conversely the product is determined by b and t.

Theorem 16.1. *If we identify* $\mathbf{R}^3$ *with the space of diagonal matrices*

$$diag(\lambda_1, \lambda_2, \lambda_3)$$

and the space V_i *with the space of Cayley numbers* x_i *(as in Theorem 15.14), then the vector space* $\mathbf{R}^3 + V_1 + V_2 + V_3$ *constructed in Theorem 14.18 is identified with* J *so that the functions* ℓ, b *and* t *correspond.*

Proof. Clearly $\ell = \lambda_1 + \lambda_2 + \lambda_3$ corresponds. Let

$$A = \begin{pmatrix} \lambda_1 & x_3 & \bar{x}_2 \\ \bar{x}_3 & \lambda_2 & x_1 \\ x_2 & \bar{x}_1 & \lambda_3 \end{pmatrix}, \qquad B = \begin{pmatrix} \mu_1 & y_3 & \bar{y}_2 \\ \bar{y}_3 & \mu_2 & y_1 \\ y_2 & \bar{y}_1 & \mu_3 \end{pmatrix}$$

then

$$2A \circ B = \begin{pmatrix} \begin{matrix} \lambda_1\mu_1 + x_3\bar{y}_3 + \bar{x}_2 y_2 \\ +\mu_1\lambda_1 + y_3\bar{x}_3 + \bar{y}_2 x_2 \end{matrix} & \begin{matrix} \lambda_1 y_3 + x_3\mu_2 + \bar{x}_2\bar{y}_1 \\ +\mu_1 x_3 + y_3\lambda_2 + \bar{y}_2\bar{x}_1 \end{matrix} & \begin{matrix} \lambda_1\bar{y}_2 + x_3 y_1 + \bar{x}_2\mu_3 \\ +\mu_1\bar{x}_2 + y_3 x_1 + \bar{y}_2\lambda_3 \end{matrix} \\ \\ \begin{matrix} \bar{x}_3\mu_1 + \lambda_2\bar{y}_3 + x_1 y_2 \\ +\bar{y}_3\lambda_1 + \mu_2\bar{x}_3 + y_1 x_2 \end{matrix} & \begin{matrix} \bar{x}_3 y_3 + \lambda_2\mu_2 + x_1\bar{y}_1 \\ +\bar{y}_3 x_3 + \mu_2\lambda_2 + y_1\bar{x}_1 \end{matrix} & \begin{matrix} \bar{x}_3\bar{y}_2 + \lambda_2 y_1 + x_1\mu_3 \\ +\bar{y}_3\bar{x}_2 + \mu_2 x_1 + y_1\lambda_3 \end{matrix} \\ \\ \begin{matrix} x_2\mu_1 + \bar{x}_1\bar{y}_3 + \lambda_3 y_2 \\ +y_2\lambda_1 + \bar{y}_1\bar{x}_3 + \mu_3 x_2 \end{matrix} & \begin{matrix} x_2 y_3 + \bar{x}_1\mu_2 + \lambda_3\bar{y}_1 \\ +y_2 x_3 + \bar{y}_1\lambda_2 + \mu_3\bar{x}_1 \end{matrix} & \begin{matrix} x_2\bar{y}_2 + \bar{x}_1 y_1 + \lambda_3\mu_3 \\ +y_2\bar{x}_2 + \bar{y}_1 x_1 + \mu_3\lambda_3 \end{matrix} \end{pmatrix}$$

Using Lemma 15.9(v), we see that the diagonal terms of $A \circ B$ are $\lambda_1\mu_1 + (x_2, y_2) + (x_3, y_3)$, $\lambda_2\mu_2 + (x_3, y_3) + (x_1, y_1)$ and $\lambda_3\mu_3 + (x_1, y_1) + (x_2, y_2)$. Thus

$$\ell(A \circ B) = \big((\lambda_1, \lambda_2, \lambda_3), (\mu_1, \mu_2, \mu_2)\big) + 2(x_1, y_1) + 2(x_2, y_2) + 2(x_3, y_3)$$

as in Theorem 14.18, so that b corresponds.

The off diagonal entries of $A \circ B$ are

$$\begin{aligned} \text{“}x_1\text{”} &: \tfrac{1}{2}(\mu_2 + \mu_3)x_1 + \tfrac{1}{2}(\lambda_2 + \lambda_3)y_1 + \tfrac{1}{2}\overline{x_2 y_3 + y_2 x_3}, \\ \text{“}x_2\text{”} &: \tfrac{1}{2}(\mu_3 + \mu_1)x_2 + \tfrac{1}{2}(\lambda_3 + \lambda_1)y_2 + \tfrac{1}{2}\overline{x_3 y_1 + y_3 x_1}, \\ \text{“}x_3\text{”} &: \tfrac{1}{2}(\mu_1 + \mu_2)x_3 + \tfrac{1}{2}(\lambda_1 + \lambda_2)y_3 + \tfrac{1}{2}\overline{x_1 y_2 + y_1 x_2}. \end{aligned}$$

Now let $C = \begin{pmatrix} \nu_1 & z_3 & \bar{z}_2 \\ \bar{z}_3 & \nu_2 & z_1 \\ z_2 & \bar{z}_1 & \nu_3 \end{pmatrix}$, and consider

$$\begin{aligned} t(A,B,C) &= b(A \circ B, C) \\ &= \lambda_1\mu_1\nu_1 + (x_1,y_1)(\nu_2+\nu_3) + (x_1,z_1)(\mu_2+\mu_3) + (y_1,z_1)(\lambda_2+\lambda_3) \\ &\quad + \lambda_2\mu_2\nu_2 + (x_2,y_2)(\nu_3+\nu_1) + (x_2,z_2)(\mu_3+\mu_1) + (y_2,z_2)(\lambda_3+\lambda_1) \\ &\quad + \lambda_3\mu_3\nu_3 + (x_3,y_3)(\nu_1+\nu_2) + (x_3,z_3)(\mu_1+\mu_2) + (y_3,z_3)(\lambda_1+\lambda_2) \\ &\quad + Re((x_2y_3+y_2x_3)z_1 + (x_3y_1+y_3x_1)z_2 + (x_1y_2+y_1x_2)z_3). \end{aligned}$$

Using Corollary 15.12(i), we see that the last line is

$$Re\Big((x_1y_2)z_3 + (y_1x_2)z_3 + (y_1z_2)x_3 + (z_1y_2)x_3 + (x_1z_2)y_3 + (z_1x_2)y_3\Big).$$

Finally, using the identification of the trilinear function $\Delta^+ \otimes \Delta^- \otimes \lambda^1 \to \mathbf{R}$ with $Re(xy)z)$ given in Theorem 15.14, our t corresponds exactly to the t in Theorem 14.18 on F_4 and it is also now clear that the function t is symmetric. □

Corollary 16.2. *The 26-dimensional representation U of F_4 is isomorphic to the subspace $\ell = 0$ of J in such a way that the products correspond up to scalars.*

Proof. This follows from Theorem 16.1 and Theorem 14.18. □

Corollary 16.3. *The group F_4 acts on J as a group of automorphisms preserving the product.*

Proof. Let F_4 act so as to fix 1 and act on the subspace $\ell = 0$ as it does on U. It then preserves ℓ and it preserves b, t on $U \otimes U$, $U \otimes U \otimes U$. Therefore F_4 preserves the component of the product from $U \otimes U$ into U, but also from $U \otimes U$ to $\mathbf{R}1$, since this is determined by $\ell(A \circ B) = b(A,B)$ and b is preserved, so it preserves products on $U \otimes U$. But certainly, it preserves products by 1, because 1 acts as a unit, so it preserves all products in J. □

So we have a monomorphism $F_4 \to Aut(J)$.

Proposition 16.4. *Any element of J can be mapped by an element of F_4 into diagonal form $diag(\lambda_1, \lambda_2, \lambda_3)$.*

Proof. Consider the elements $A = \begin{pmatrix} \lambda_1 & x_3 & \bar{x}_2 \\ \bar{x}_3 & \lambda_2 & x_1 \\ x_2 & \bar{x}_1 & \lambda_3 \end{pmatrix}$ in an F_4-orbit. The function $||x_1||^2+||x_2||^2+||x_3||^2$ is continuous on this orbit and the orbit is compact because F_4 is compact. Therefore the function attains its minimum in the orbit. Suppose, for a contradiction, that the minimum is strictly greater than zero and attained at some element of the orbit with $x_1 \neq 0$. F_4 contains three copies of $Spin(9)$, one for each of V_1, V_2, V_3. The one corresponding to V_1 acts according to the representation λ_9^1 on the 9-dimensional space spanned by V_1 and matrices $diag(0, \lambda, -\lambda)$. It acts by Δ on $V_2 + V_3$, the space of the x_2 and x_3. $Spin(9)$ is transitive on vectors of a fixed length in λ_9^1, so by using an element of $Spin(9)$, we can replace our representative by one with a new λ_2, λ_3 and new $x_1 = 0$. This does not change $||x_2||^2 + ||x_3||^2$, so it diminishes $||x_1||^2 + ||x_2||^2 + ||x_3||^2$, contradicting our assumption that our representative attained the minimum value. □

Addendum 16.5. *A diagonal matrix $diag(\lambda_1, \lambda_2, \lambda_3)$ can have its diagonal entries permuted by elements of F_4.*

Proof. By using an element of $Spin(9)$ ($\subset F_4$) we can preserve $diag(1,0,0)$ and $diag(0,1,1)$ and reverse $diag(0,1,-1)$. That is, we can interchange λ_2, λ_3. Similarly we get the other transpositions and hence all of Σ_3. □

Theorem 16.6. (i)
The algebra of polynomial functions on J invariant under $Aut(J)$

= the algebra of polynomial functions on J invariant under F_4

= the polynomial algebra generated by the functions $\ell(A)$, $b(A,A)$, $t(A,A,A)$.

(ii) *An element of J is determined up to action of F_4 by the values of $\ell(A)$, $b(A,A)$ and $t(A,A,A)$ there.*

(iii) *Any element of J satisfies a cubic equation*

$$A^3 - c_1 A^2 + c_2 A - c_3 I = 0,$$

where $A^2 = A \circ A$, $A^3 = (A \circ A) \circ A = A \circ (A \circ A)$, $c_1 = \ell(A)$, $c_2 = \frac{1}{2}(\ell(A)^2 - b(A,A))$ and $c_3 = \frac{1}{6}(\ell(A)^3 - 3\ell(A)b(A,A) + 2t(A,A,A))$. (Note that the product $\circ$ is commutative, but not associative!)

Proof. (i) Call the three algebras P_1, P_2, P_3, respectively. Already we have $P_3 \subset P_1 \subset P_2$. A polynomial function of J, invariant under F_4, is determined by its values on diagonal matrices $diag(\lambda, \mu, \nu)$, by Proposition 16.4. The value is a polynomial in λ, μ, ν and symmetric by Addendum 16.5. Thus it is a polynomial in the elementary symmetric functions or in the power sums $\lambda+\mu+\nu$, $\lambda^2+\mu^2+\nu^2$, $\lambda^3+\mu^3+\nu^3$, which are the restrictions of $\ell(A)$, $b(A,A)$, $t(A,A,A)$ respectively to diagonal matrices. Hence $P_2 \subset P_3$ and (i) is proved.

(iii) It is sufficient to check for a real matrix $A = diag(\lambda, \mu, \nu)$. For such a matrix we get $A^3 - c_1A^2 + c_2A - c_3 = 0$, where $c_1 = \lambda+\mu+\nu$, $c_2 = \lambda\mu+\mu\nu+\nu\lambda$, $c_3 = \lambda\mu\nu$.

(ii) Any element of J can be mapped by F_4 onto $diag(\lambda, \mu, \nu)$ by Proposition 16.4. The functions $\ell(A)$, $b(A,A)$, $t(A,A,A)$ determine λ, μ, ν up to order (by part (iii)) which determines $diag(\lambda, \mu, \nu)$ up to the action of F_4 by Addendum 16.5. □

Theorem 16.7. (i) *$F_4 \cong Aut(J)$ by the above action.*

(ii) *The subgroup of $Aut(J)$ fixing $e_1 = diag(1,0,0)$ is $Spin(9)$.*

(iii) *The subgroup of $Aut(J)$ fixing e_1, $e_2 = diag(0,1,0)$ and $e_3 = diag(0,0,1)$ is $Spin(8)$.*

Proof. (iii) Consider the linear map $J \to J$ given by $A \mapsto e_1 \circ A$. Its 1-eigenspace is 1-dimensional, generated by e_1. Its 0-eigenspace is 10-dimensional, generated by e_2, e_3 and V_1, the space of matrices in which the first row and column are zero. Its $\frac{1}{2}$-eigenspace, of dimension 16, is $V_2 + V_3$, the space of matrices of the form $\begin{pmatrix} 0 & x_3 & \bar{x}_2 \\ \bar{x}_3 & 0 & 0 \\ x_2 & 0 & 0 \end{pmatrix}$ If $\theta : J \to J$ fixes

e_1, it preserves $V_2 + V_3$,

e_2, it preserves $V_3 + V_1$,

e_3, it preserves $V_1 + V_2$.

Thus, if θ preserves e_1, e_2 and e_3, it preserves V_1, V_2 and V_3 separately and such a θ consists of linear maps $1 : \mathbf{R}^3 \to \mathbf{R}^3$, $V_i \to V_i$, preserving inner products and f. Thus θ lies in $Spin(8)$ by Lemma 15.15.

(ii) Let $\theta : J \to J$ preserve e_1. It also preserves $1 = e_1 + e_2 + e_3$, therefore preserves $e_2 + e_3$. Now $e_1 \circ (e_2 - e_3) = 0$, so $\theta(e_1 \circ (e_2 - e_3)) = \theta e_1 \circ \theta(e_2 - e_3) = e_1 \circ \theta(e_2 - e_3) = 0$. Thus $\theta(e_2 - e_3)$ is in the 0-eigenspace of $A \mapsto e_1 \circ A$. We conclude that $\theta(e_2 - e_3) = \begin{pmatrix} 0 & 0 & 0 \\ 0 & \lambda & x_1 \\ 0 & \bar{x}_1 & -\lambda \end{pmatrix}$, where $\lambda^2 + \bar{x}_1 x_1 = 2$. So there exists $g \in Spin(9)$ such that $g(e_2 - e_3) = \theta(e_2 - e_3)$ and we can reduce the problem to the case of $\phi = g^{-1}\theta$ fixing e_1, e_2, e_3. Thus $\phi = g^{-1}\theta \in Spin(8)$ by (iii) and $\theta \in Spin(9)$.

(i) Suppose $\theta : J \to J$ is an automorphism and consider θe_1. This is a point A with $\ell(A) = 1 = b(A,A) = t(A,A,A)$. By Theorem 16.6(ii), there exists $g \in F_4$ such that $g^{-1}\theta e_1 = e_1$. Thus we can reduce the problem to the case of an automorphism $\phi = g^{-1}\theta$ which preserves e_1; such a ϕ must lie in $Spin(9)$ by part (ii). Hence $\theta \in F_4$. □

The Cayley projective plane.

In constructing this plane, our problem is, of course, the lack of associativity. We try to learn from the case of $\mathbf{H}$. A point of $P^2(\mathbf{H})$ is a 1-dimensional submodule $\subset \mathbf{H}^3$ and this can be identified with a projection operator, namely "orthogonal projection onto this 1-dimensional subspace", which is a 3×3 Hermitian matrix, A, over $\mathbf{H}$, idempotent ($A^2 = A$) and of rank 1.

In the exceptional Jordan algebra of 3×3 matrices over $\mathbf{O}$, let Π be the F_4-orbit of $e_1 = diag(1,0,0)$, that is, the space of all A such that $\ell(A) = 1$, $b(A,A) = 1$, $t(A,A,A) = 1$: such A *do* satisfy $A^2 = A$ and have rank 1. Hence $\Pi = F_4/Spin(9)$ is a *Cayley projective plane.* (Cf. Theorem 16.7 (ii).) Recall from the proof of Theorem 16.7 that e_1, acting on J by $\circ$, has a

1 eigenspace of dimension 1 generated by e_1;
$\frac{1}{2}$ eigenspace of dimension 16, the tangent space, $V_2 + V_3$, to Π at e_1;
0 eigenspace of dimension 10, generated by e_2, e_3 and V_1.

The points of Π in the 0 eigenspace, i.e. the rank 1 idempotents orthogonal to e_1, form an S^8 with $Spin(9)$ acting on it — the space of points in Π which attain maximum distance from e_1. They constitute a Cayley line in Π. (See [He, p. 167].) As befits a projective plane, two distinct points of Π determine a unique Cayley line joining them (and two distinct Cayley lines meet in a unique point of Π). The condition for three points to be collinear is the following. Take the function $\lambda\mu\nu$ corresponding to $c_3 = \frac{1}{6}\big(\ell(A)^3 - 3\ell(A)b(A,A) + 2t(A,A,A)\big)$, and polarise to get the associated symmetric trilinear function

$$\ell(A)\ell(B)\ell(C) - \ell(A)b(B,C) - \ell(B)b(C,A) - \ell(C)b(A,B) + 2t(A,B,C).$$

This function is ≥ 0 on Π and is equal to zero if and only if A, B, C are collinear. Finally, by transposing this trilinear function, we get

$$A \times B = 2A \circ B - \ell(A)B - \ell(B)A + (\ell(A)\ell(B) - b(A,B))I,$$

a point on the Cayley projective plane determining the equation of the line joining A and B, viz., the set of points which attain maximum distance from $A \times B$ or are orthogonal to $A \times B$.

Appendix. Jordan algebras

Definition. *A* special Jordan algebra *is a subset of an associative algebra which is closed under the new product* $A \circ B = \frac{1}{2}(AB + BA)$.

Examples. In $M_n(\mathbf{F})$ with $\mathbf{F} = \mathbf{R}, \mathbf{C}$ or $\mathbf{H}$, take the subset of Hermitian matrices.

Note. *The class of special Jordan algebras cannot be characterised by algebraic identities.*

Proof. There is a homomorphism from a special Jordan algebra onto an algebra which is not a special Jordan algebra. For details see [Co]. □

In a special Jordan algebra, we have the following

Proposition. (i) $(A \circ B) \circ (A \circ A) = A \circ (B \circ (A \circ A))$.

(ii) *The associator, as a function of its middle variables, is a derivation with respect to* $\circ$, *that is* $[A, B \circ C, D] = [A, B, D] \circ C + B \circ [A, C, D]$.

Proof. (i) $(A \circ B) \circ (A \circ A)$

$$= \tfrac{1}{2}(AB + BA) \circ A^2$$
$$= \tfrac{1}{4}(A^3B + ABA^2 + A^2BA + BA^3)$$
$$= \tfrac{1}{4}(A^3B + A^2BA + ABA^2 + BA^3)$$
$$= \tfrac{1}{4}(A(BA^2 + A^2B) + (BA^2 + A^2B)A).$$

(ii) $[A, X, D]$

$$= (A \circ X) \circ D - A \circ (X \circ D)$$
$$= \tfrac{1}{2}((AX + XA) \circ D - A \circ (XD + DX))$$
$$= \tfrac{1}{4}(AXD + XAD + DAX + DXA - AXD - ADX - XDA - DXA)$$
$$= \tfrac{1}{4}(\,[X, AD] - [X, DA]\,)$$
$$= \tfrac{1}{4}(\,[[D, A], X]\,).$$

But in any associative algebra,

$$\begin{aligned} [E, XY] &= EXY - XYE \\ &= EXY - XEY + XEY - XYE \\ &= [E, X]Y + X[E, Y], \end{aligned}$$

that is, the commutator is a derivation as a function of the second variable. So,

$$\begin{aligned} [E, X \circ Y] &= \tfrac{1}{2}(\,[E, X]Y + X[E, Y] + [E, Y]X + X[E, Y]\,) \\ &= [E, X] \circ Y + X \circ [E, Y]. \end{aligned}$$

□

In fact in a special Jordan algebra, (i) $\Rightarrow$ (ii), but (ii) $\not\Rightarrow$ (i). The classical definition of a *Jordan algebra* is an algebra with a commutative product $\circ$, such that (i) $(A \circ B) \circ (A \circ A) = A \circ (B \circ (A \circ A))$ holds.

A Jordan algebra which is not special is called *exceptional*. For example the algebra J of Chapter 16 is exceptional.

References

[1] J. F. Adams, *Lectures on Lie groups,* Benjamin, 1969; reprint by the University of Chicago Press, 1982.

[2] M. F. Atiyah, R. Bott and A. Shapiro, *Clifford Modules,* Topology, 3 Supplement 1 (1964), pp. 3–38.

[3] N. Bourbaki, *Eléments de mathématique: Groupes et algèbres de Lie,* Chap. 6, Hermann, 1975.

[4] N. Bourbaki, *Groupes et algèbres de Lie,* Chap. 3, Hermann, 1972.

[5] E. Cartan, *Leçons sur la théorie des spinors II,* Hermann, 1938, pp. 53–54.

[6] E. Cartan, *Le principle de dualité et la théorie des groupes simples et semi-simples,* Bull. des Sci. Math., 49 (1925), pp. 361–374.

[7] C. Chevalley, *Theory of Lie groups,* Princeton University Press, 1946.

[8] B. Eckmann, *Gruppentheoretischer Beweis der Satzes von Hurwitz-Radon uber die Komposition quadratischer Formen,* Comment. Math. Helv., 15 (1942), pp. 358–366.

[9] S. Helgason, *Differential geometry and symmetric spaces,* Academic Press, 1962.

[10] N. Jacobson, *Lie algebras,* John Wiley, 1967.

[11] I. Porteus, *Topological geometry,* Van Nostrand Reinhold Co., 1969.

[12] H. Weyl, *The classical groups,* Chap. 5, Princeton University Press, 1946.

ADDITIONAL REFERENCES

[Ad 1] J. F. Adams, *2-Tori in* E_8, Math. Ann., 287 (1987), pp. 29–39.

[Ad 2] J. F. Adams, *The fundamental representations of* E_8, Contemp. Math., 37 (1985), pp. 1–10.

[Ad 3] J. F. Adams, *Spin*(8), *triality,* F_4 *and all that,* in Superspace and supergravity, ed. S. W. Hawking and M. Rocek, Cambridge University Press, 1981, pp. 435–445 .

[Al] A. A. Albert, *A structure theory for Jordan algebras,* Ann. of Math., 48 (1947), pp. 546–567, especially p. 550.

[All] B.N.Allison, *A class of nonassociative algebras with involution containing the class of Jordan algebras,* Math. Annalen, 237, no.2, pp. 133–156 (1987).

[Br] R. Brauer and H. Weyl, *Spinors in n dimensions,* Amer. J. Math., 57 (1935), pp. 425–449.

[Bro] R. B. Brown, *Groups of type* E_7, J. f. reine u. angew. Math. 236(1969), pp. 79–102.

[Co] P. M. Cohn, *Universal algebra,* D. Reidel Publishing Co., 1982, p. 308.

[Fr] H. Freundenthal, *Oktaven, Ausnahmegruppen und Oktavengeometrie,* Geometriae Dedicata, 19 (1985), pp. 1–63.

[Ha] Melvin Hausner and Jacob T. Schwartz, *Lie Groups and Lie algebras,* Gordon and Breach, Science Publishers, Inc. 1968.

[He] S. Helgason, *Groups and geometric analysis: integral geometry, invariant differential operators and spherical functions,* Academic Press Inc., 1984.

[La] S. Lang, *Algebra*, Addison Wesley Pub. Co. Inc., 1965, Chapter 16.

[S] Edwin H. Spanier, *Algebraic Toplogy*, Springer-Verlag, New York Inc, 1989.